Free-standing Tension Structures

Free-standing Tension Structures

From tensegrity systems to cable-strut systems

Wang Bin Bing

CRC Press
Taylor & Francis Group
Boca Raton London New York

CRC Press is an imprint of the
Taylor & Francis Group, an **informa** business
A SPON PRESS BOOK

CRC Press
Taylor & Francis Group
6000 Broken Sound Parkway NW, Suite 300
Boca Raton, FL 33487-2742

First issued in paperback 2019

ISBN-13: 978-0-415-33595-9 (hbk)
ISBN-13: 978-0-367-86550-4 (pbk)

Visit the Taylor & Francis Web site at
http://www.taylorandfrancis.com

and the CRC Press Web site at
http://www.crcpress.com

Typeset in Sabon by
Newgen Imaging Systems (P) Ltd, Chennai, India

British Library Cataloguing in Publication Data
A catalogue record for this book is available from the British Library

Library of Congress Cataloging in Publication Data
A catalog record for this book has been requested

Contents

Figures

Tables

Preface

How to obtain 'non-column' space rationally and beautifully has been a dream of mankind for years. It is always our ambition to develop elegant, fascinating, unbelievable and even 'unreasonable' and 'impossible' forms. There seems no limit to architectural art. Meanwhile, a designer has to face a combination of different objectives and constraints such as safety, costs, aesthetics, manufacturing and functional requirements. Normally, rationality in structural point of view is required.

Tension structures extend architects' imagination due to cables' lightness, unlimited length and flexibility, strengthening the thought that the structure itself be exposed as a part of aesthetics rather than be enclosed. The shapes of polyhedral space frames have interested people for years. How to adapt polyhedra to our architectural and structural needs invites innovation in forms.

Tensegrity (tensile integrity) systems, characteristic of free-standing nature and isolation of struts in cable networks, have become a popular topic in recent years among structural engineers, architects, researchers in the world of space structures under the driving forces of pursuing lightness and innovation in forms as is the tendency of the contemporary research. As new structural forms, tensegrity grids are composite forms of tension structures and pin-jointed space frames. The book presents the load-transfer properties inherent in tensegrity systems and extends the concept to a wider scope – cable-strut systems with structurally more efficient grid types by arming the law of mechanics and with a broadened expression of polyhedral art in architecture and special functions.

The book contains two parts. Part I includes the concept and properties of tensegrity systems:

Chapter 1: Background of tension structures and introduction to tensegrity systems.

Chapter 2: Properties of tensegrity simplexes, background knowledge and general analytical method for mechanism-contained pin-jointed systems.

Chapter 3: Structural configurations, design and structural properties of tensegrity grids, structural principles on how to invent new systems with much-improved structural efficiency.

Part II is focused on the improved cable-strut systems:

Chapter 4: Cable-strut concept and geometrical characterization of simplexes and structural configurations.
Chapter 5: Mechanical properties, design principles and design examples of lightweight cable-strut grids.
Chapter 6: Further application information, including large span design, joint design and deployable design.
Chapter 7: Architectural aspect of cable-strut systems presented with new concepts and a wide variety of forms.

The book is the first effort to provide an in-depth investigation of tensegrity system from structural point of view and also the first to study how to stabilize polyhedra by struts and cables and use them as building blocks to form structurally efficient grids with rich results. The effort lies in three aspects. The book presents *structural mechanics* generally suitable for all pin-jointed systems with simplified way of analysis. It provides the generalized method of analysing pin-jointed systems with mechanisms (Chapter 2), the principle of how to form structurally efficient free-standing tension systems based on the law of mechanics (Chapter 3), the application of the principle to stabilize basic polyhedra as building blocks leading to a wide variety of new cable-strut grids (Chapter 4), and mechanical analysis exposing their properties by plane analogue truss method (Chapter 5). It appeals generally to technical-minded readers or students with interest in extensive bar-system structural mechanics.

In view of architecture, the book gives architects more choices to express architectural art by structural forms. It presents a wide variety of basic structural forms (Chapters 3 and 4) and developed forms (Chapter 7) concerned with architecture. Based on the principle, more forms made of basic and higher polyhedra can be developed from and beyond what are described in the book, enriching the application and study of polyhedra architecture. The inherent merits in deployable and retractable functions (Chapter 6) extend the application to special functions. *In view of engineering applications*, innovative structural forms presented in the book (Chapters 3 and 4) provide new solutions in engineering practice. The provided design principle and examples (Chapters 3, 5 and 6) act as the tool for structural design. The simplicity in joint design (Chapter 6) is also presented. The information in the book is sufficient for marketing these products. Both aspects appeal to architects, engineers, researchers, construction companies and graduate students (including final-year undergraduates) having interest in the innovation of tension structures and space frames.

The author wishes to develop curious and rational structural forms to serve society. Generally, the topic of the study is relatively new and the work is aimed at inspiring new thoughts in the field and inviting exploration, discovery, and application of these systems to creative, imaginative, expressive and potential designs for buildings. If the book contains anything that deserves the term 'creative,' the author owes a debt of gratitude to all our great predecessors for showing the way.

Wang Bin Bing
Singapore, 2003

Acknowledgements

I can only think warmly about one person. He is my supervisor, Prof. Liu XiLiang, in Tianjin University, PRC. During my PhD studies, Prof. Liu always encouraged me to develop on my own. I was deeply moved, as this is not a usual occurrence in China. Especially at the dinner celebrating my debate of PhD thesis, my supervisor asked me to sit at his usual post. It was for the first time and the last. The scene is rooted in my heart and the words are still ringing by my ear, 'jie jiao, jie zao'. After all these years, I still continue to feel that there is a pair of kind eyes full of expectation looking at me.

On 9 January 1995, I submitted my first paper to the *International Journal of Space Structures*. At that time, I was a PhD student in Tianjin University, PRC. On 16 January, Prof. H. Nooshin, the chief editor in University of Surrey, UK gave me a warm reply and tried to find a patient and experienced reviewer to improve the standard of the paper. The paper was poorly written, especially in language and style, but it was a crucial start, especially for a young man who was then ignorant of the outside

Prof. Liu XiLiang (left) and the author (Photo taken: Ms Li YanYun).

Tensegrity model made by the author (Wang 1995).

world. Later I met Prof. Nooshin several times at international conferences. Every time, I received kind suggestions and blessings from him.

It is beyond words to express my gratitude to Dr Ariel Hanaor, of Israel Institute of Technology. Ariel reviewed a large part of my journal papers and especially the manuscript of this book. I should say, without Ariel's time and patience, my study could not have developed so quickly. As an acknowledged pioneer in the field, Ariel never hesitated to contribute his own idea to enhance the technical and language level of my work. Paper by paper, chapter by chapter, with lots of time and energy, with the pioneer's blessing, the book is being published!

I owe my thanks to Mr Matthys Levy, in Weidlinger Associates, USA, from whom I learned a lot about cable domes. I cannot forget Prof. Lewis Schmidt in the University of Wollongong, for his kind blessing and help. I should thank Prof. Richard Liew for kind suggestions and care in the National University of Singapore. I also should thank Prof. Rene Motro in the University of Montpellier, who gave me valuable suggestions on papers. I am grateful to Mr W. Hughes from Multi-Science Publishing and *Journal of IASS* for kindly granting free quotation of materials. Special gratitude is owed to *Newgen Imaging Systems (P) Ltd*, for ensuring figure quality of this book.

The book summarizes my personal work incorporating background knowledge, which was accomplished entirely in my leisure time ever since my graduation in April 1996 (I also made a tensegrity model myself (see above) during the PhD study, which attracted lots of people's attention when it was made). This was made possible by the lady who has accompanied me all these years, pulling through adversities together, taking on the responsibilities of the whole family so that I could concentrate my mind on something 'impractical'. She is the lady behind – Ms Li YanYun. Moreover, as the co-author of lots of papers, my wife contributed significantly to this book with structural analysis coupled with lots of fancy drawings and ideas.

Finally, I believe that the book will encourage my daughter, HuiYuan, to persist in her wish.

Wang Bin Bing

Part I

Tensegrity systems

1 Introduction

In the recent years, space structures as a whole are becoming lighter, and even lightness has become a fashion in aesthetics. A common phenomenon is that tension structures are becoming more and more popular as these structures extend architects' imagination due to cables' lightness, unlimited length and versatility. The combination of tension structures with roof material, such as glass and membrane, expresses the sense of transparency of space and lightness of forms.

Another driving force is the innovation in forms. Forms keep changing so as to achieve pleasant looking and unique structures to meet the curiosity of people, especially when the forms themselves are also a type of tension structures. The typical case is polyhedral space frames.

In space structures architectural art is often expressed by the structure itself, hence new structural forms create new aesthetic standards. As a type of tension structures, tensegrity systems studied in this book are typical examples and is a popular topic today. However, as originality and personal expression are their first concern, architects may not pay much attention to structuralism, or rationality in structural principle. The study on structural rationality leads to the extension of tensegrity systems to cable-strut systems.

In this chapter, the application of tension structures is reviewed. In view of definition, there seems to be some controversies on what are tension structures. Some scholars consider that membrane is a part of tension structures. Some others consider it only as a roof material. The discussion in this book is based on the latter. Followed by the background review, tensegrity concept is introduced as the extensive application of tension structures.

1.1 Classification of space structures

Space structures of spanning types can be classified into three basic categories according to the dominant load-transfer patterns (Figure 1.1):

- Catenary-like types (Figure 1.1(a)), in which the dominant load-transfer pattern is the axial tension. Equilibrium of the structure is obtained by the compression sustained by the boundary anchoring system or

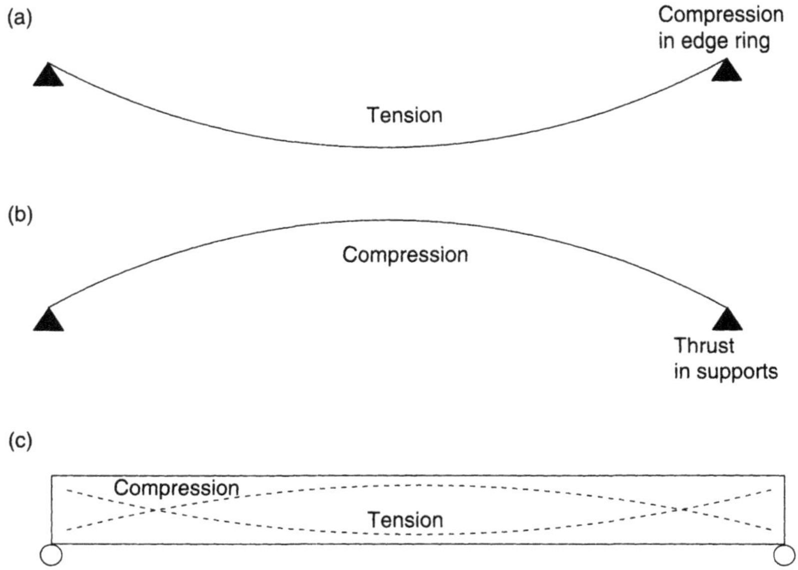

Figure 1.1 Classification of space structures: (a) catenary-like type; (b) arch-like type; (c) beam-like type (dashed line: dominant force flow).

supports. The structure is not free-standing and consequently, load response is transferred to the boundary. Cable-suspended forms are the most familiar of these types.

- Arch-like types (Figure 1.1(b)), in which the dominant load-transfer pattern is the axial compression. Thrust forces are balanced by supports. Usually, the structure itself can be free-standing under its own weight but the deformation and especially self-weight may be much larger without supports. Familiar forms include various forms of latticed or continuum shells.
- Beam-like types (Figure 1.1(c)), in which the dominant load-transfer pattern is a self-equilibrated combination of compression and tension. The structure is free-standing. The global internal forces are identified as bending moments and shears at the section. Space frames are the popular of these types.

1.2 Roles of cables in space structures

This section summarizes the basic roles cables playing in three basic structural types. Other applications may be feasible in the combined types. Meanwhile, some milestone projects are mentioned here so as to demonstrate the application and development of tension structures.

1.2.1 *Catenary-like types*

In catenary-like types, cables always act as the principal structural components due to their natural advantages in sustaining tensions (Figure 1.2).

Pure cable net forms

In 'pure' cable net forms, only cables are structural components. In order to apply prestress and avoid the water drainage problem, special shape, such

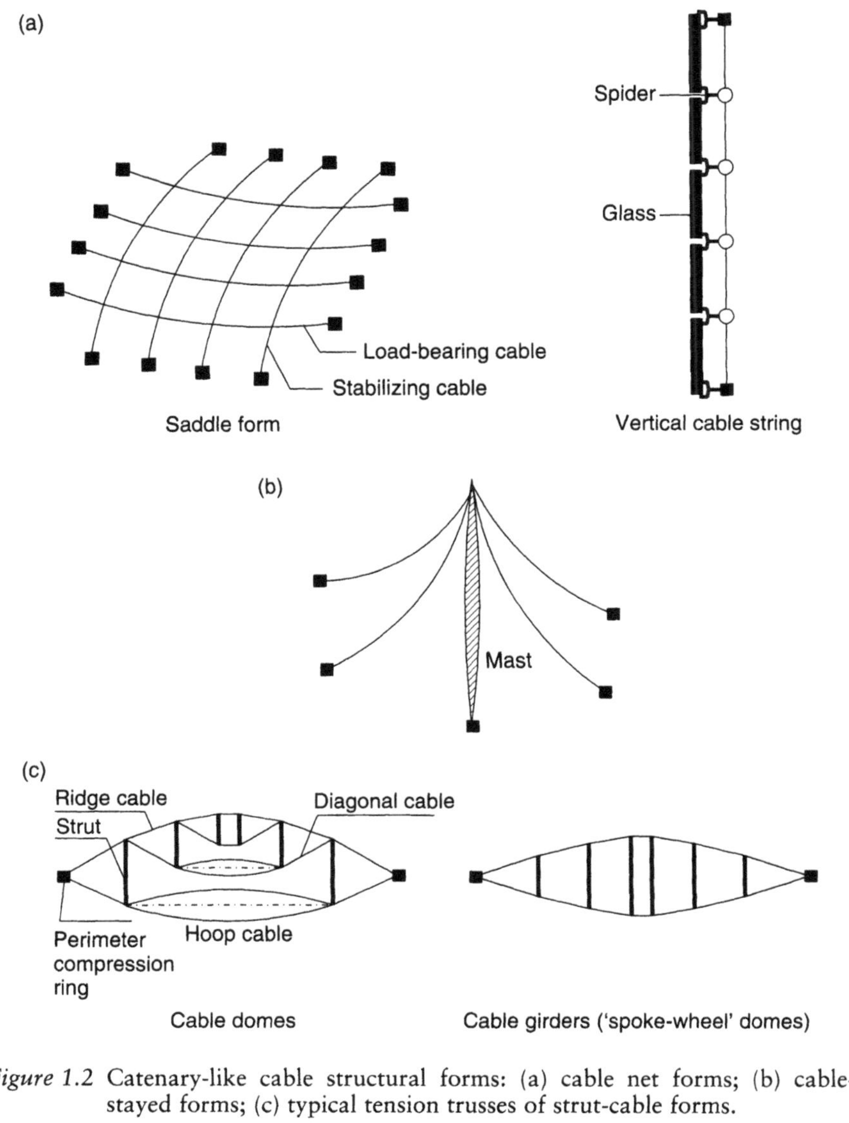

Figure 1.2 Catenary-like cable structural forms: (a) cable net forms; (b) cable-stayed forms; (c) typical tension trusses of strut-cable forms.

as saddle (Figure 1.2(a)), has to be designed. In addition, some planar cable net form is often used in vertical cladding for glass curtain walls (Figure 1.2(a)).

Cable-stayed forms

Cable-stayed forms are similar to cable net forms. The difference is that cable-stayed forms are supported by bulky masts and the ground in place of a bulky ring beam (Figure 1.2(b)). It affords architects more freedom in imagination.

Cable-stayed forms can be developed in a linear or circular way. In the linear way, the basic unit is copied consecutively (Figure 1.3). Notable projects include the New Denver International Airport in US, measuring approximately 90m by 300m in plan, which is supported by 34 masts of roughly 30m in length (Brown 1998).

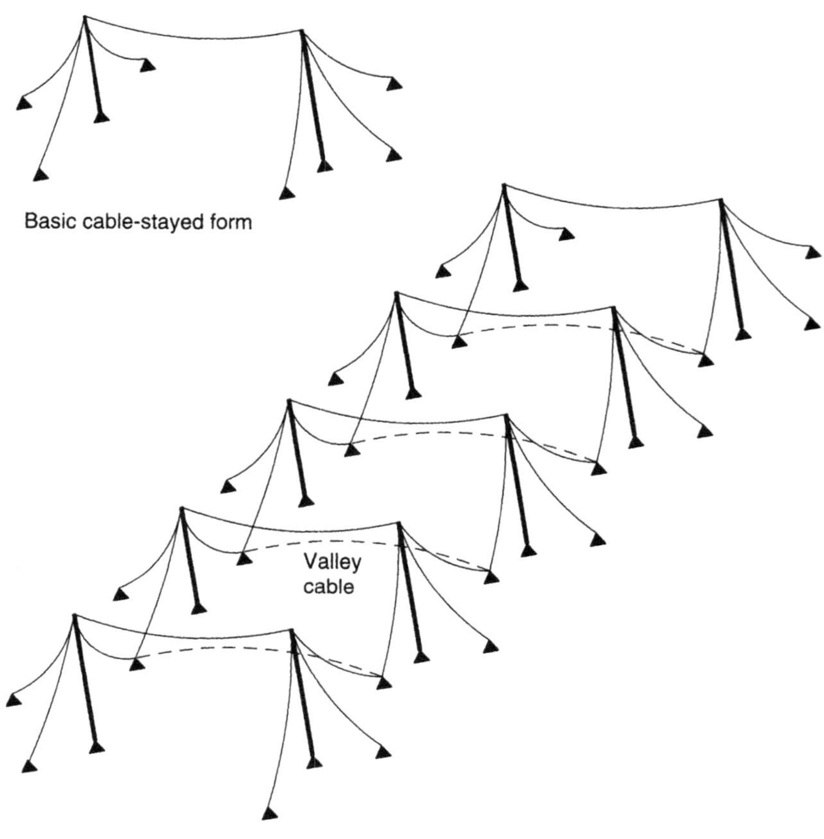

Figure 1.3 Linear arrays of basic cable-stayed forms.

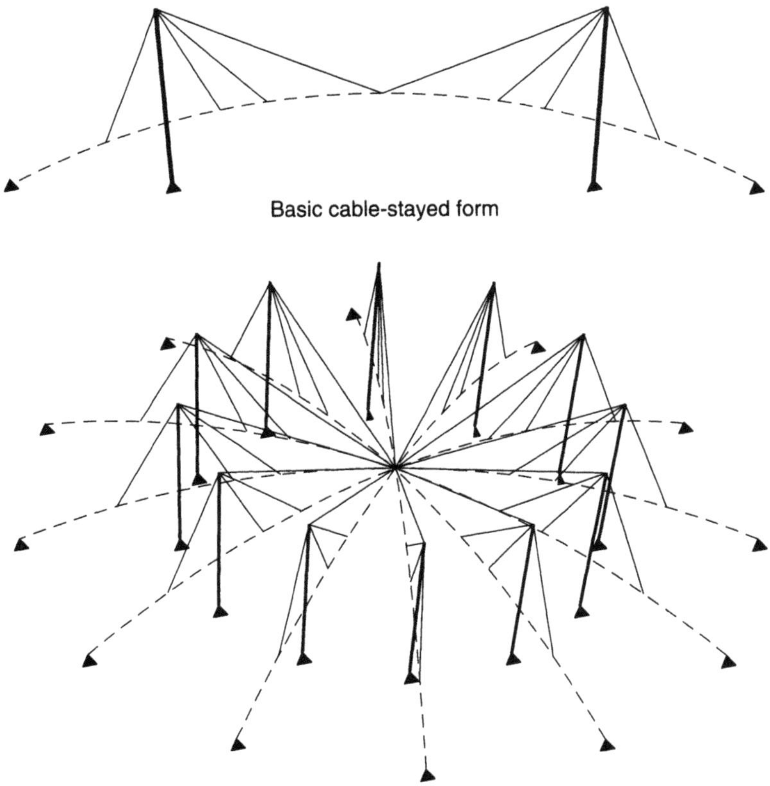

Figure 1.4 Circular arrays of basic cable-stayed forms.

In the circular way, the basic unit is copied radially (Figure 1.4). Notable projects include the Millennium Dome in UK <http:// www.burohappold. com>. It covers 80,000m^2, containing 12 huge masts, which support long radial cables reaching 150m from the perimeter to the centre (Liddell (1998)). As circular form, opening can be designed so as to show different look (Figure 1.5).

Strut-cable net forms

In strut-cable net forms, isolated struts are introduced into cable nets to form the outwardly curved surface. Basic forms include cable girders ('spoke-wheel' domes) and cable domes. Strut-cable net forms generally follow truss-like behaviour (Hanaor 2002), and their typical tension trusses are given in Figure 1.2(c). The spatial applications of these trusses render various forms.

Cable domes are relatively new between two basic strut-cable net forms. A cable dome contains vertical struts, ridge cables, diagonal cables and

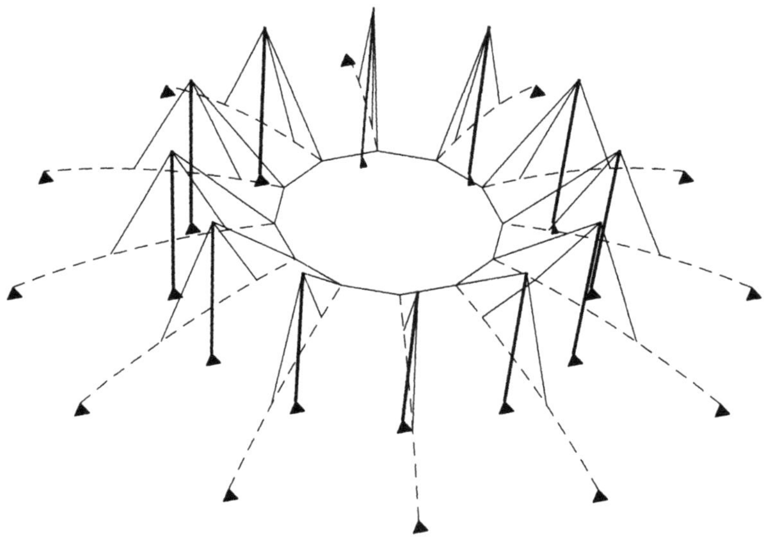

Figure 1.5 Cable-stayed forms with opening.

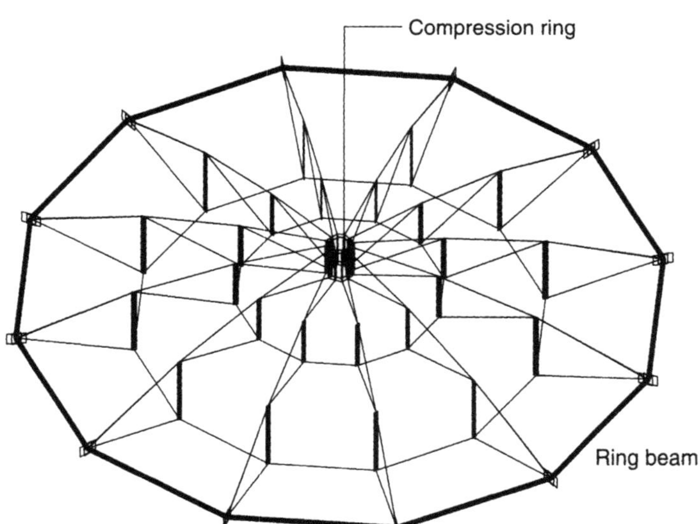

Figure 1.6 Geiger's dome.

hoop cables (Figure 1.2(c)). Cable domes mainly include two variations: Geiger's domes (Figure 1.6), and spatially triangulated domes (Figure 1.7). In the Geiger's dome, cable trusses are arranged radially. The first cable dome of the type is the Gymnastics Arenas for the 1986 Korean Olympics

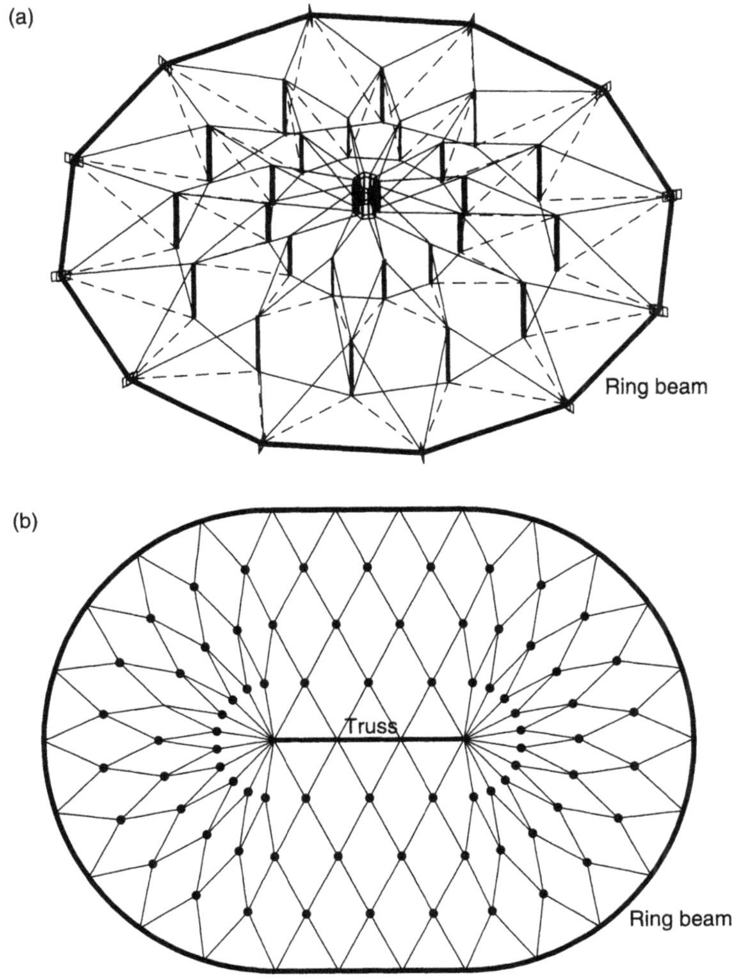

Figure 1.7 Spatially triangulated domes: (a) circular form; (b) elliptical form (top layer).

<http://www.geigerengineers.com>. In the spatially triangulated dome, adjacent rings of struts offset half a module. The Georgia Dome, measuring 240m in length and 193m in short, is among very few of the kind constructed today <http://www.wai.com> (Levy 1994, Brown 1998). It is roughly an elliptical shape with a truss linking two focuses (similar to Figure 1.7(b)), with an incredibly low weight of $30 kg/m^2$ based on the high strength of cables.

Cable girders can be used to cover full roof plan (Figure 1.8(a)). In addition, it has the varied cable wheel forms in which a big opening is designed at the centre (Figure 1.8(b)). An example is the Outdoor Stadium Roof for

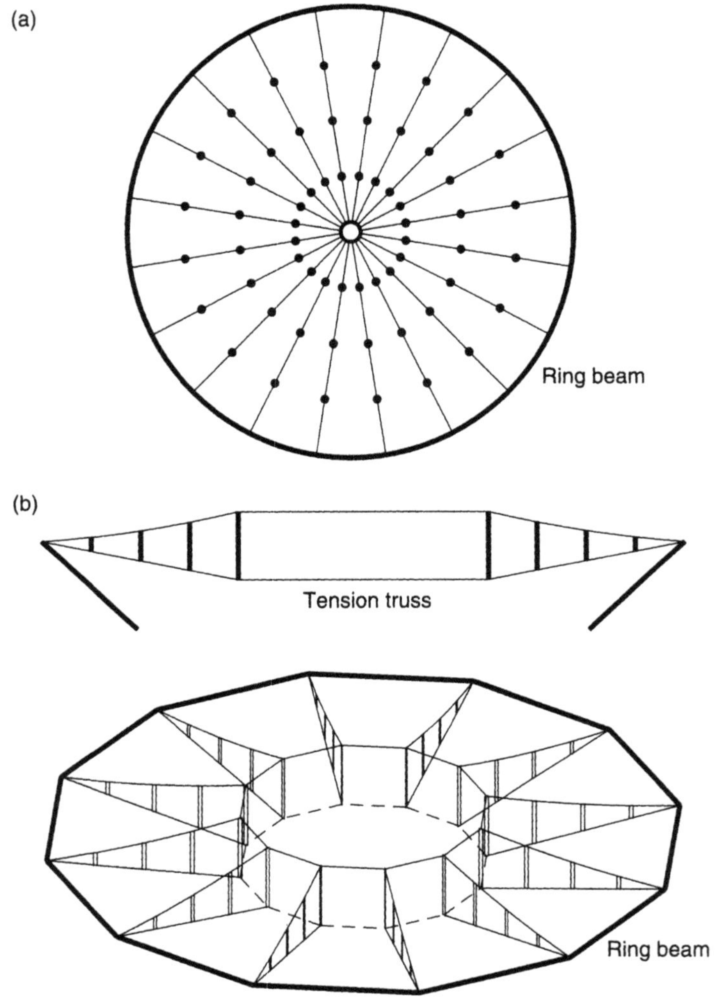

Figure 1.8 Cable girder domes: (a) full-plan form; (b) form with opening.

the National Sports Complex in Kuala Lumpur (Schlaigh 1997). Moreover, it is worthy to point out that the evolved planar form of 'spoke-wheel' type is often used for supporting glass (Figure 1.9).

1.2.2 *Arch-like types*

Due to their weakness in sustaining compression forces, cables are most frequently used as reinforcing or stabilizing components in arch-like structures, rather than as structural components like bars.

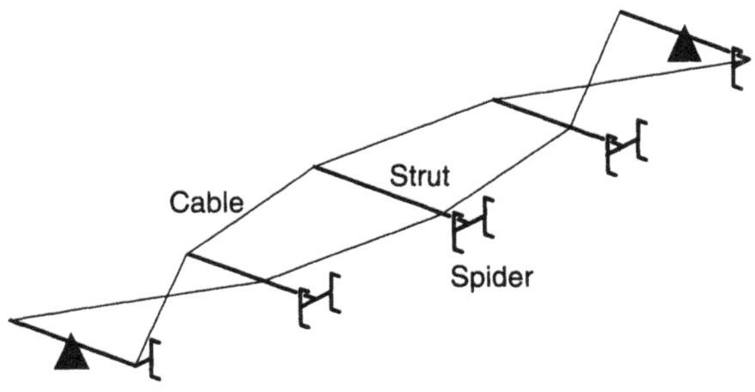

Figure 1.9 Skeleton for glass.

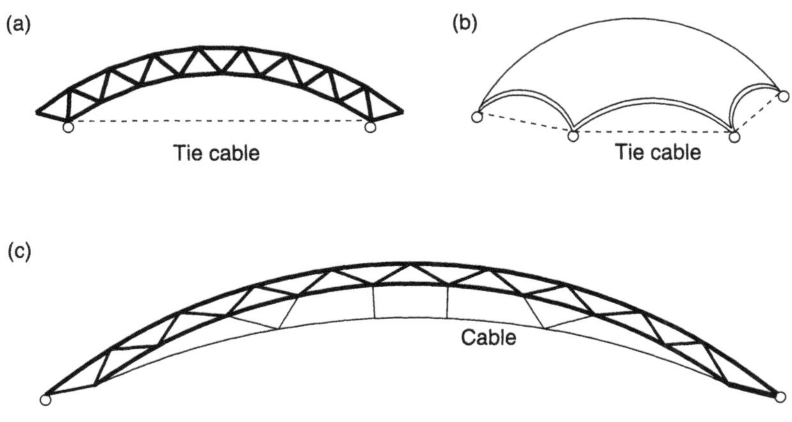

Figure 1.10 Reinforcing arch-like forms by cables: (a) tie cables in cylindrical shells; (b) tie cables in domes; (c) reinforcing cables in arch.

Reinforcing components

As reinforcing components, cables are used to improve the stiffness and force distribution, and/or reduce thrust forces in supports. Typical examples are tie cables in supports in cylindrical shells (Figure 1.10(a)) or domes (Figure 1.10(b)). Cables can also be introduced into the arch directly (Figure 1.10(c)). One such example of combining arches and cables is the UNI-dome in the University of Northern Iowa (Berger 1997, Pilla 1998).

The other aspect of applications is to reinforce arch-like types by the assembly of struts and cables. The resulting forms include hyper cable domes or hyper 'spoke-wheel' domes, which can also be considered as the composite forms of arch-like types and cable domes or 'spoke-wheel' domes. The design of hyper cable domes is illustrated in Figure 1.11, including radial form and spatially triangulated form. The ring beam can be avoided but

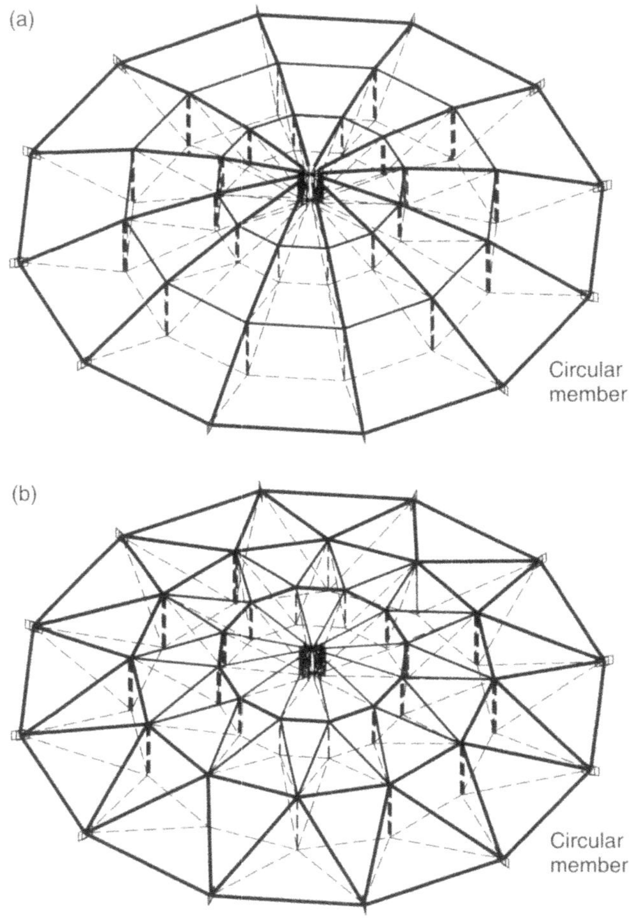

Figure 1.11 Hyper cable domes: (a) radial form; (b) spatially triangulated form.

circular members are still indispensable. Examples include the 'tension-brace dome structure' of Marin Midland Arena (radial form) (Brown 1998), and the 'cable-suspen' dome (spatially triangulated form) (Kawaguchi *et al.* 1999). These forms are basically of the same principle and the arch benefits from the introduction of strut-cable assembly at the improvement on the overall stiffness (stability) and the reduction in thrust forces to circular components under gravity load.

Stabilizing components

Cables can also be used to stabilize the building units, which are otherwise deformable. A typical example is the grid shell (Figure 1.12) (Schlaigh and

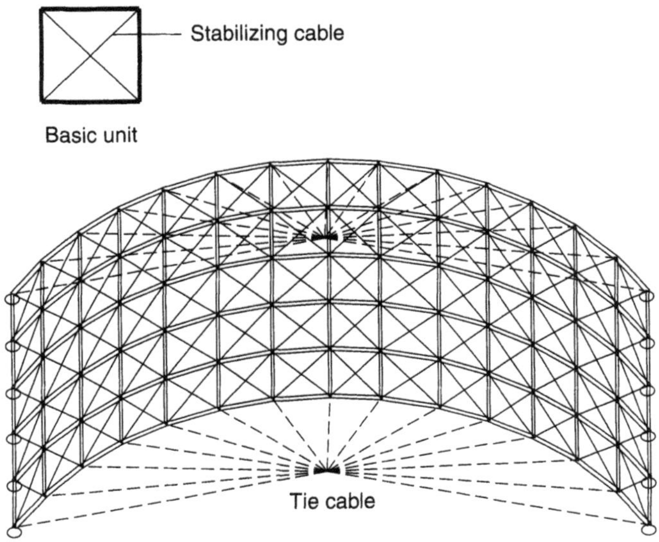

Figure 1.12 A grid shell form.

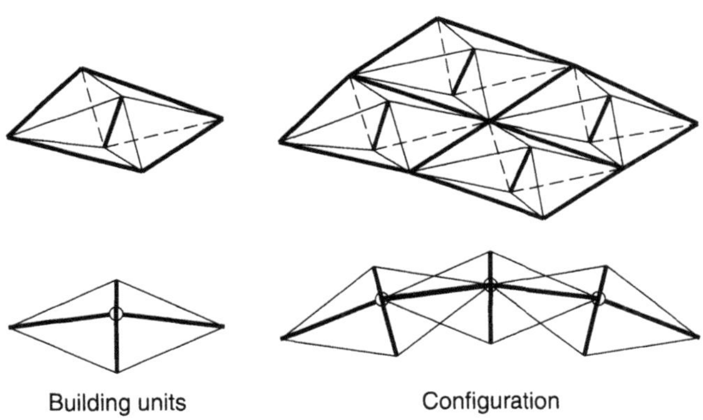

Figure 1.13 Unit structure (Adapted from Hangai *et al.* 1992).

Schober 1999) in which two crossed cables are introduced in each rectangular panel enclosed by steel beams (note that the tie cable assembly of spider form is also introduced as reinforcing components at a certain interval along generatrix). The design avoids using the conventional triangular panels. The grid shell may have varied forms so as to improve the bending strength (Saitoh 1998).

Another example is the unit structure (Figure 1.13) developed by Hangai *et al.* (1992). In the building unit, the enclosed four hinged struts in the base are stabilized by a strut and eight cables with prestress. The base of the unit is not co-planar so as to reserve curvature in the resulting shell forms.

Moreover, among new forms developed in this book that prestress is not essential for sability, cables can stabilize one layer of struts (certain discussion in Section 6.1.3) or can be sandwiched between two or three layers of struts (Section 4.6).

1.2.3 Beam-like types

Cables serve as principal structural components in catenary-like types and as stabilizing or reinforcing components in arch-like types. In comparison, cables in beam-like types can be used as structural components together with bars so as to form new tension structures in addition to their natural advantage as stabilizing or reinforcing components.

Stabilizing or reinforcing components

As stabilizing components, cables can replace web chords in space truss grids to stabilize each building block (Figure 1.14). These cables can be applied either at the face or inside each building block.

As an example of reinforcing components, cables are introduced in the tensional layer (e.g. the bottom layer when downward load is dominant) of

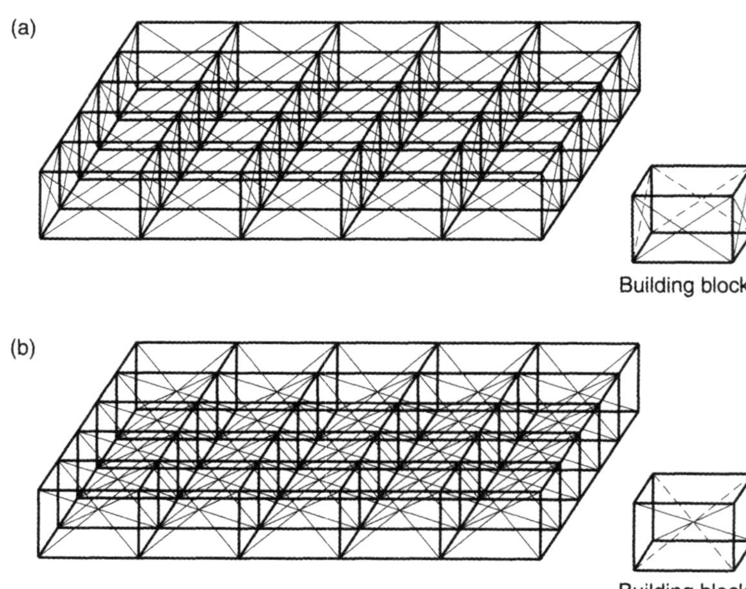

Figure 1.14 Cables as stabilizing components in space trusses: (a) cables at the faces of building blocks; (b) cables inside building blocks.

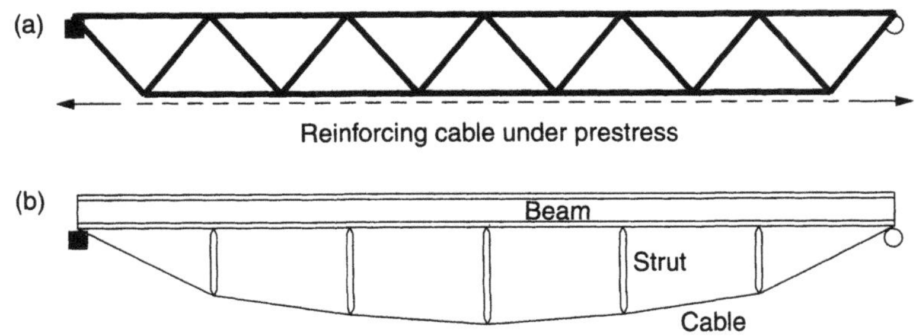

Figure 1.15 Cables as reinforcing components: (a) in space trusses; (b) under beam with struts.

space bar grids (Figure 1.15(a)). Prestress is introduced in cables to improve force distribution with some weight reduction. Another example, the assembly of struts and cables can be applied under the beam to improve stiffness and strength under gravity load (Figure 1.15(b)).

Structural components

As structural components, cables in beam-like forms sustain tensional components of internal moments, forming free-standing cable networks. Tensegrity grids and new cable-strut grids with extended concept belong to the type. These systems broaden the application scale of tension structures but are actually also new types of space frames (so they can also be used in arch-like forms in which cables mainly serve as stabilizing components whereas bars form the load-carrying skeleton, as mentioned in Section 1.2.2). Tensegrity concept is introduced in the following section.

1.3 Introduction to tensegrity

'Tensegrity' attracted wide attention since its mysterious origin and its unique characteristics in architecture. Although some similar ideas occurred occasionally in 1920s (Motro 1996), topic studies were led by Buckminister Fuller, the originator of geodesic geometry. He pushed forward the idea and coined the word 'tensegrity' from 'tensile integrity'. Kenneth Snelson materialized the idea into prototypes. However, first true tensegrity unit was filed by David Emmerich in 1960s.

1.3.1 *Origin of tensegrity*

Fuller's conviction

In the 1940s, Buckminister Fuller, a well-known architect, was convinced that the universe operated according to a tensional integrity principle. This belief gave birth to 'tensegrity' a mysterious background. As reported by Sadao (1996): 'To Fuller, tensegrity is Nature's grand structural strategy. At the cosmic level, he saw that the spherical astro-islands of compression of the solar system are continuously controlled in their progressive repositioning in respect to one another by comprehensive tension of the system which Newton called "gravity". At the atomic level, man's probing within the atom disclosed the same bind of discontinuous compression, continuous tension apparently governing the atom's structure.'

Fuller believed that tensegrity principle is the rule of the nature from the solar system to the atom system that is tied by gravity forces. But his conviction neglected that these systems follow '*kinematic equilibrium*', substantially different from '*static equilibrium*' relation in actual construction structures. Despite that, the conviction still led to the invention of new structural and architectural systems.

Snelson's models

Fuller did not manage to materialize his confound convictions himself. Later in the summers of 1947 and 1948, Fuller taught at Black Mountain College and spoke constantly of 'tensile integrity'. He hoped to create a model of his structural principle.

In late 1948, Kenneth Snelson, a student and now sculptor, presented three models to Fuller (Snelson 1996). Among them, the last one – X-column piece (Figure 1.16) contained the preliminary idea of 'continuous tension, discontinuous compression'. It is composed of wooden cross with wires stretched in a square across the tips of cross. The two units were then placed across each other in a three-dimensional way and were attached to a piece of wooden plate. According to Snelson's description, a 'planar' tensegrity module is obtained. It contains two bars and four cables. Cables are independent in sustaining internal forces. But they are linked at ends and thus are 'continuous' in view of geometry. Two bars connect only to cables at ends. They look crossing each other but do not contact at body (although it is not possible in two-dimensional geometry). It is the tensions that tie the system together as one integrity.

Later, Snelson continued his effort to separate successfully compression members at the points where they cross each other by introducing cables in a three-dimensional way. He developed a number of such sculptures based on these improved 'X' modules. The most famous one is the 'Needle Tower' at the Hirshhorn Museum of Modern Art, Washington, DC. (Figure 1.17), as only independent struts could be seen from afar. These ornamental forms demonstrate 'islands of compression in a sea of tension' fascinatingly. So some people believe that Snelson is the originator of tensegrity. However, it is interesting to note that the sculptor himself did not believe tensegrity is prospering as load-bearing structures (Snelson, personal communication).

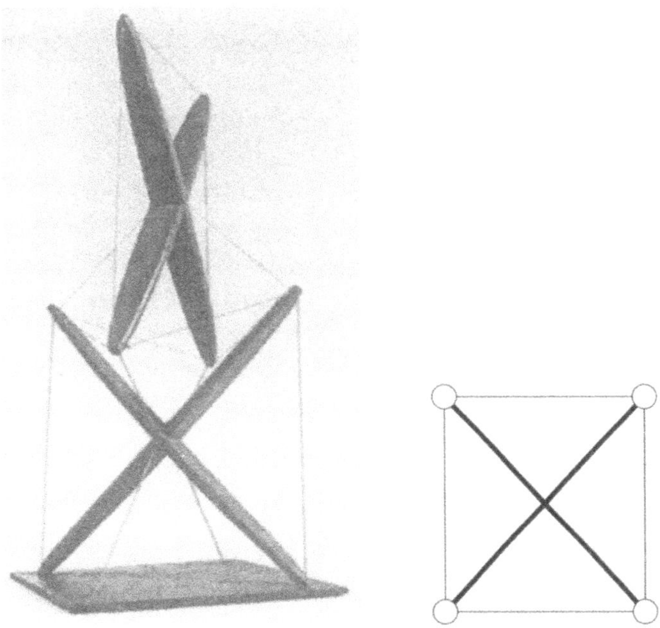

Figure 1.16 Snelson's X module (left), and the 'planar' tensegrity simplex derived from the module (right).

Source: Snelson (1996); Courtesy: Multi-Science Publishing.

Fuller's idea

Return to our story, when he caught sight of Snelson's model, Fuller believed that it was what he searched for years, and his idea of tensegrity became concrete. He coined the word 'tensegrity' in the 1950s, and defined the notions of tensegrity as follows (Fuller 1975):

> The word tensegrity is an invention: it is a contraction of tensile integrity. Tensegrity is guaranteed by the finitely closed, comprehensively continuous, tensional behaviours of the system and not by the discontinuous and exclusively local compressional member behaviours. ... The integrity of the whole structure is invested in the finitely closed, tensional-embracement network, and the compressions are local islands. ... Tension is omni-directionally coherent. Tensegrity is an inherently non-redundant confluence of optimum structural effort effectiveness factors. All structures, properly understood, from the solar system to the atom, are tensegrity structures. Universe is omni-tensional integrity.

Due to Fuller's personal influence, people took the assertion for granted and believed that tensegrity was inherently structurally efficient for years.

Figure 1.17 Snelson's 'Needle Tower' at the Hirshhorn Museum of Modern Art, Washington, DC (Photo courtesy: Hanaor).

Fuller (1996) went on developing tensegrity structural forms, but unfortunately, these forms suffer from various kinds of structural problems. A typical example is his patent of 'tensile integrity structure' (Fuller 1962), whose load-carrying capacity relies on the buckling resistance of crossed long bars and whose bars are constructed as flexible tensegrity truss in very long spans (Figure 1.18). It is the invention of the simplest three-dimensional tensegrity object – tensegrity simplexes that opened wide research field on tensegrity structures.

1.3.2 Tensegrity simplexes

Tensegrity simplexes are the elementary spatial tensegrity system. In fact, Snelson's free-standing assembly of X modules can be evolved from the

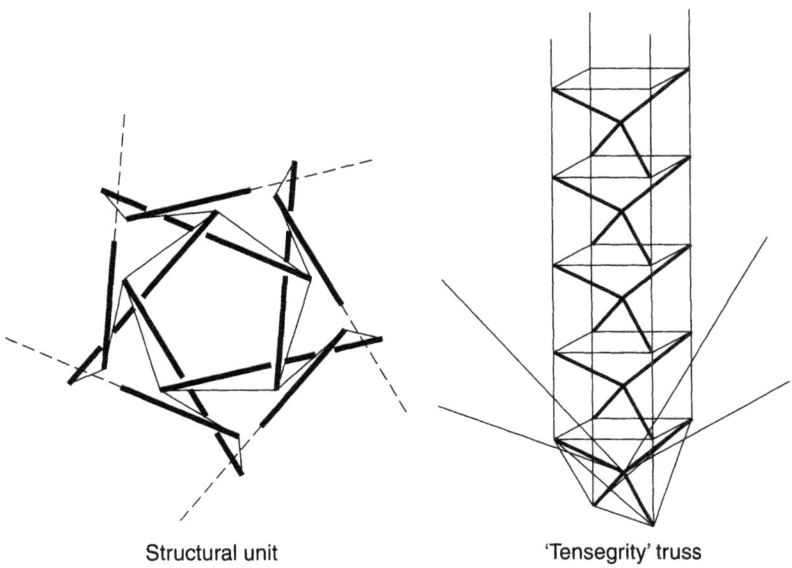

Figure 1.18 Fuller's patent of 'tensile integrity structure'.
Source: Fuller (1962).

array of some tensegrity simplexes. Originally, tensegrity simplexes refer to the triangular tensegrity prism. Later, the concept is extended to higher prisms and their varied forms.

A tensegrity prism is a stable (free-standing) volume that is realized by the interaction between isolated struts (islands of compression) and interconnected cables (a sea of tension). It is an anti-prism (occasionally, prism) composed of two layers of cables forming the upper base (by upper cables) and the bottom base (by bottom cables), stabilized by diagonal cables. Inclined struts connect opposite vertices of bases so as to brace the prism (Figure 1.19(a)). The relative rotation angle of the two bases is dictated by the requirement for the equilibrium of the shape, as explained later in Section 2.1. For regular simplexes (i.e. having regular base polygons), this angle is 30° for a triangular anti-prism, and 45° for a square anti-prism, etc. (Chassagnoux *et al.* 1992).

Tensegrity prisms appeared first in David Emmerich's patent (Emmerich 1963). In fact, he found it in 1958 through independent research (Emmerich 1996). Interestingly, he used the term 'self-tensioning structures' to define his idea and never adopted the term 'tensegrity', although they are essentially the same.

In a tensegrity prism, if the sizes of the upper base and bottom base are designed not identical, a tensegrity truncated pyramid is formed (Figure 1.19(b)), which is basically of the same mechanical properties. Various forms of tensegrity simplexes allow the forming of a wide variety of tensegrity structures, which is discussed in Chapter 3.

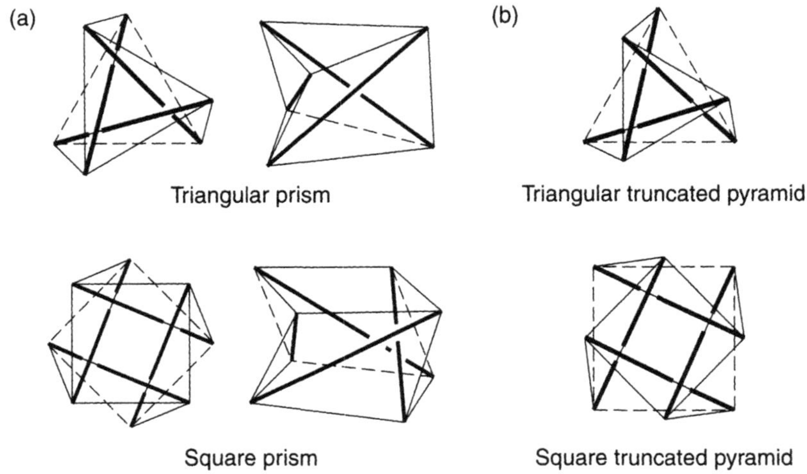

Figure 1.19 Tensegrity simplexes: (a) tensegrity prisms; (b) tensegrity truncated pyramids.

1.3.3 *Higher tensegrity polyhedra*

In addition to the invention of tensegrity prisms, Emmerich also studied how to form higher tensegrity polyhedra of the same principle, which was called by him 'single-layer spherical structures' (Emmerich 1990). He studied all Platonic polyhedra (regular polyhedra) and Archimedian polyhedra (semi-regular polyhedra), whose vertices are inscribed into a sphere. All these polyhedra are convex since the tensional members must be on the outside, farther from the centre of the polyhedron than the compression members that are inserted without any contact among themselves. Among five Platonic polyhedra, the tetrahedron is unrealizable; the octahedron and the cube are variants of the triangular and square simplexes, respectively; the icosahedron and the dodecahedron are given in Figure 1.20(a). Among tensegrity Archimedians, the simplest forms, cuboctahedron and pseudo cuboctahedron are given in Figure 1.20(b). For high tensegrity Archimedians refer to Emmerich (1990).

Another way to form tensegrity polyhedra is proposed by Grip (1992). All the vertices of Platonic polyhedra are truncated (but not at the centre of the edges). Each original vertex is thus transformed into three or four vertices. Struts are inserted to connect each pair of new vertices that are transformed from originally adjacent vertices but are not in the same face. Tensegrity truncated tetrahedron, truncated cube, truncated octahedron are presented in Figure 1.21. It seems not simple to connect higher tensegrity polyhedra to form structural grids. However, these polyhedra enrich the application of tensegrity principle in architectural aspect.

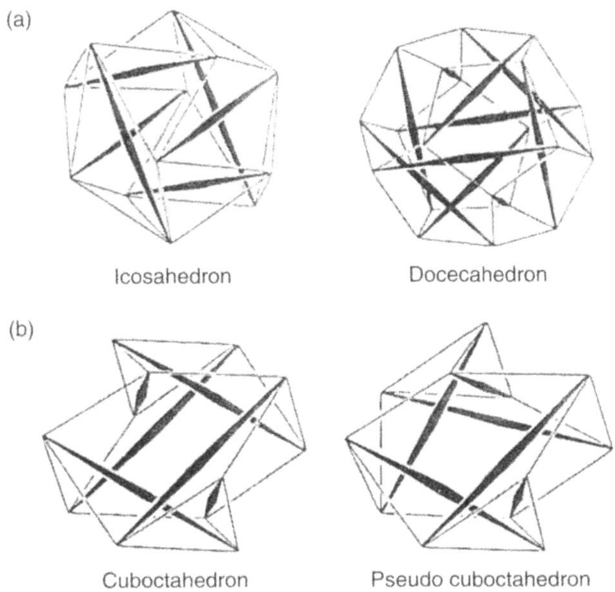

Figure 1.20 Higher tensegrity polyhedra: (a) platonic tensegrity polyhedra; (b) tensegrity Archimedians.

Source: Emmerich (1990); Courtesy: Multi-Science Publishing.

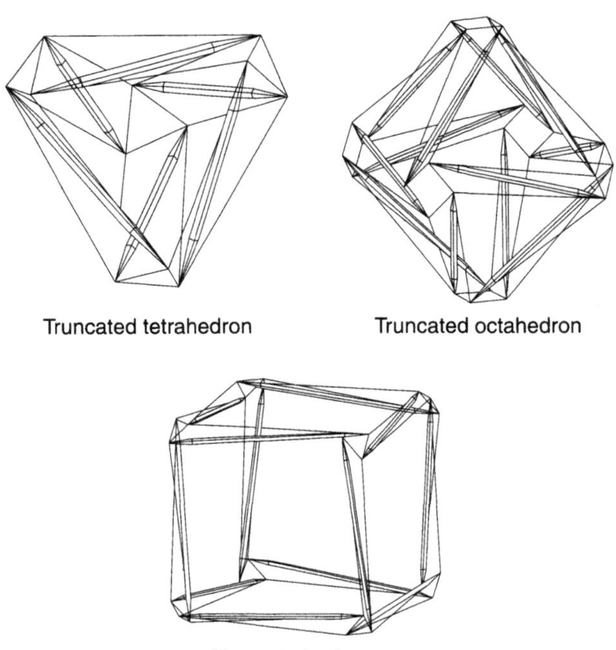

Figure 1.21 Truncated tensegrity polyhedra.

Source: Grip (1992); Courtesy: Multi-Science Publishing.

1.3.4 *Structural definition of tensegrity systems*

Currently, the definition of tensegrity is subjected to controversies partially due to a large variety of forms developed. All these controversies ignored, tensegrity systems, as demonstrated by tensegrity prisms and higher tensegrity polyhedra, can be defined from structural point of view:

> Tensegrity systems are free-standing pin-jointed networks in which an interconnected systems of cables are stressed against a disconnected system of struts or extensively, any free-standing systems composed of tensegrity units satisfying the aforesaid definition.

In the definition, the former half includes 'pure' tensegrity forms, whereas the latter half is more general with extended concept.

Theoretically, struts are recti-linear components that have lower bound on length but can be extended freely in a 'telescoping' way and therefore they do not sustain tension. In structural use, struts are straight bars that are designed to sustain only compression (e.g. in a tensegrity polyhedra, bars in tension normally does not occur if we check nodal equilibrium condition). Cables are recti-linear members with upper bound on length, but can contract freely (slacken) and therefore cannot sustain compression (unless they have pretension reserve).

The definition distinguishes tensegrity systems from conventional cable networks that rely on bulky anchoring system and prestress as the prerequisite to achieve equilibrium of the whole structure. Nevertheless, it is interesting to note that cable domes and 'spoke-wheel' domes are derived from tensegrity principle. Both types of domes span the space with continuous tensional cables and discontinuous compressive posts having the appearance of 'islands of compression in a sea of tension'. Therefore, some structural engineers take them as tensegrity systems. However, their mechanical properties are substantially different (catenary-like types vs beam-like types).

2 Properties of tensegrity simplexes and analysis of pin-jointed systems containing mechanisms

Among tensegrity polyhedra, tensegrity simplexes appear most suitable as building blocks for structural grids. Other polyhedra have complex geometry and thus it seems not easy to connect them to form grids with necessary stiffness. In this chapter, mechanical properties of tensegrity simplexes are explained in detail incorporating basic knowledge related to mechanisms. As the analytical tool, the general method of analysing pin-jointed systems containing mechanisms is presented based on the stiffness-based Newton iteration and dumb component method. Properties of the grids formed by tensegrity simplexes are discussed in the next chapter.

2.1 Equilibrium condition in tensegrity prisms

A tensegrity prism is stable only at a specific prism rotation angle (the relative rotation angle between two bases). The relation can be obtained from the nodal internal force equilibrium condition under an assumed self-stress state. In a tensegrity prism, it can be easily got from symmetrical conditions that the forces are equal for all upper and bottom cables, for all inclined cables, and for all struts, respectively. Meanwhile, the equilibrium condition for each joint is the same. Each joint is in equilibrium under the application of actions exerted by the strut, the inclined cable and the two cables in the base. The strut is confined within the three-dimensional volume enclosed by the connecting three cables, which is the minimum for equilibrium.

In a triangular prism, for example, at an arbitrary joint *j*, we can replace the forces in two cables *jk* and *jl* in the base by a resultant. The direction *jj'* of the resultant force passes through the joint *j* and the base centre O and *jj'* should be co-planar with the connected strut *jm* and the inclined cable *jn* for equilibrium (Figure 2.1). As the base cable *mn* is co-planar with the strut *jm* and the inclined cable *jn*, the direction line *jj'* should be parallel to *mn* as they are already in parallel base planes. The prism rotation angle obtained is apparently 30°, half the vertex angle.

By repeating this procedure, the prism rotation angle for the square tensegrity prism is 45°. The analysis shows the specific prism rotation angle

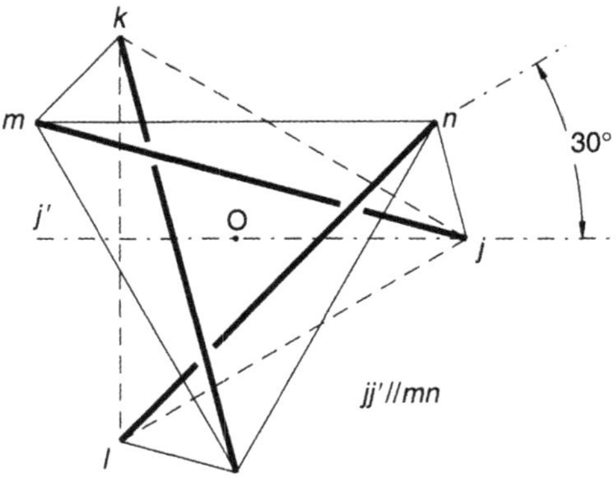

Figure 2.1 Equilibrium condition of a triangular tensegrity prism.

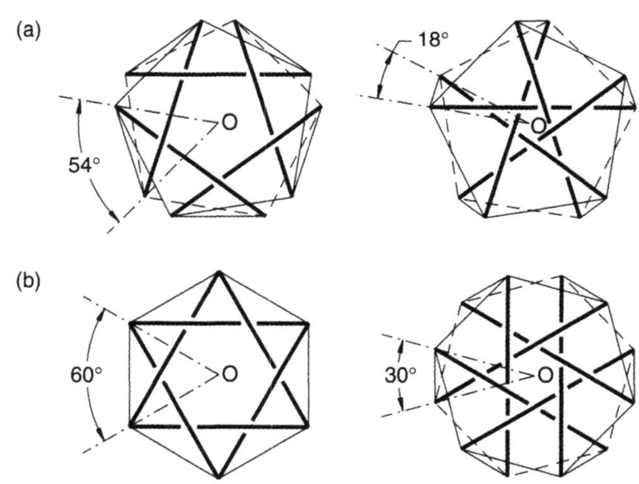

Figure 2.2 Higher tensegrity prisms with more than one prism rotation angle: (a) pentagonal prism; (b) hexagonal prism.

Source: Chassagnoux *et al.* (1992); Courtesy: Multi-Science Publishing.

for the stabilization of a tensegrity prism, which is half the vertex angle in base. As the geometrical characteristics, being larger or smaller than this value results in instability of the volume. However, for tensegrity prisms higher than the triangular and the square, other specific angles may also be found (Figure 2.2) (Chassagnoux *et al.* 1992).

The equilibrium condition also implies that each tensegrity prism has one and only one self-stress state. The self-stress state is related to another aspect of the mechanical properties of tensegrity prisms, that is, mechanism analysis, which also reveals mechanisms contained in each tensegrity prism.

2.2 Mechanism analysis of tensegrity simplexes

In this section, the mechanism analysis of tensegrity prisms is presented. As the background knowledge, classification of various pin-jointed systems on mechanisms is introduced first, followed by two analytical methods of mechanisms.

2.2.1 Classifications of pin-jointed systems

Pin-jointed systems refer to the systems in which all components are only subjected to axial forces. These components include bars that can be subjected to both compression and tension, and cables (tendons) that can only be subjected to tension. Pin-jointed systems include a large variety of space structures, such as space trusses (all bars) and cable suspended structures (all cables or dominantly cables), etc. Tensegrity prisms and the resulting structures (mixed bars and cables) obviously belong to pin-jointed systems.

The classification of pin-jointed systems is based on the independent self-stress states (s) and independent mechanisms (m). Note that the classification is based on the initial geometry and that mechanisms may be finite mechanisms or infinitesimal mechanisms. An infinitesimal mechanism disappears after a displacement less than the length order whereas a finite mechanism still exists.

A system that contains infinitesimal mechanism has at least one self-stress state. Given a system with $m > 0$ and $s > 0$, if a state of self-stress can impact positive first-order stiffness to every mechanism, then the mechanisms are first-order infinitesimal, that is, they are associated with second-order changes of component lengths. If on the other hand there are some mechanisms that cannot be stabilized by a state of self-stress, these mechanisms are (at least) second-order infinitesimal or are finite (Calladine and Pelligrino 1991). Two examples in Figure 2.3(c) and (d) illustrate respectively first- and second-order infinitesimal mechanisms. However, second- or high-order infinitesimal mechanisms seldom appear in actual structures.

Geometrically deformable pin-jointed systems that contain finite mechanisms ($s = 0$, $m > 0$) are in general not considered a structure (Figure 2.3(a), (b)). Pin-jointed systems that do not contain finite mechanisms can be classified as follows:

- When $s > 0$, $m > 0$, the system is statically and kinematically undetermined, or geometrically flexible (Figure 2.3(c), (d)).
- When $s = 0$, $m = 0$, the system is statically and kinematically determined (Figure 2.3(e)).
- When $s > 0$, $m = 0$, the system is statically undetermined and kinematically determined (Figure 2.3(f)).

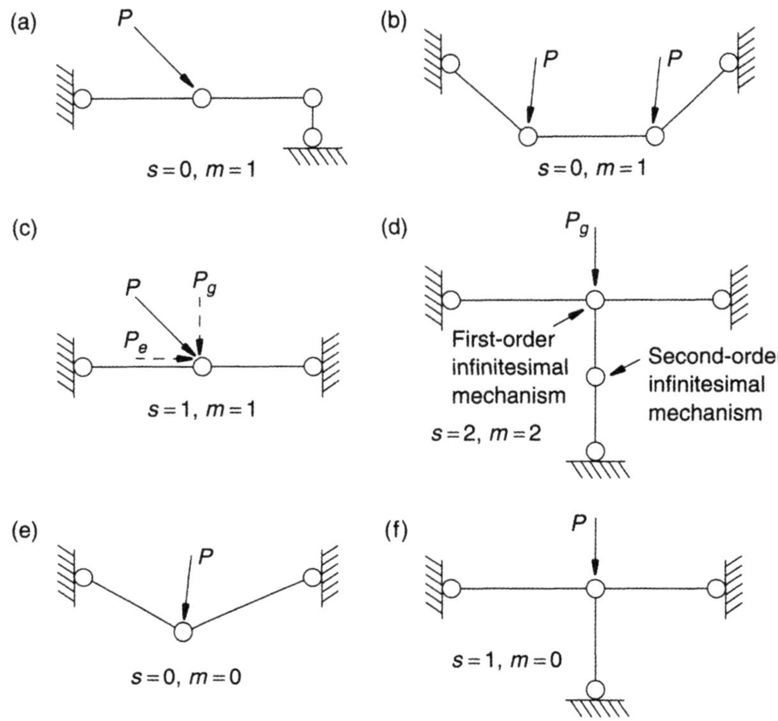

Figure 2.3 Classification of pin-jointed bar systems by two-dimensional examples: (a) and (b) geometrically deformable; (c) and (d) geometrically flexible; (e) and (f) geometrically rigid.

Geometrically flexible system cannot maintain equilibrium with the applied load to the mechanism in its original geometry (the load–deflection curve has zero slope at the origin). But it can deform and change its shape to develop internal force components to balance the external load. The magnitude of the deformation can be quite large, even for small load values. Initial stiffness has to be obtained by a state of self-stress. These properties are discussed in Section 2.4.1.

Two mechanism free cases, such as those in Figure 2.3(e) and (f) are also described as 'geometrically rigid' cases. As a geometrically rigid system does not contain mechanisms, it can maintain equilibrium in its original geometry. Its deformation is a result of elastic deformation of components and hence is normally much smaller compared with that of a geometrically flexible system, unless its geometry is near mechanism.

The characteristics of pin-jointed systems are compared in Table 2.1 for clarity.

Table 2.1 Characteristics of pin-jointed systems

Concepts	*Mechanisms contained*	*Stable?*	*Having initial stiffness under external force?*	
			Non-prestress state	*Prestressed state*
Geometrically deformable	Finite	No	No	Not applicable
Geometrically flexible	Infinitesimal	Yes	No	Yes
Geometrically rigid	No	Yes	Yes	Yes

2.2.2 *Two analytical methods of mechanisms*

Extended Maxwell's rule

As a tool to analyse the self-stressed states and mechanisms of pin-jointed bar systems, the extended Maxwell's rule is introduced here (Calladine 1978):

$$2N_j - N_e - N_c = m - s, \quad \text{for two-dimensional systems} \tag{2.1a}$$
$$3N_j - N_e - N_c = m - s, \quad \text{for three-dimensional systems} \tag{2.1b}$$

where N_j, N_e and N_c are the number of joints, elements and external constraints, respectively. In tensegrity structures, cables are allowed to slacken under load. At this state, N_e becomes the number of force-carrying elements, excluding slack cables. But in general, the analysis is based on the initial geometry.

In a pin-jointed system with relatively simple geometry, either m or s can be obtained from simple nodal equilibrium analysis, then the other can be derived from Eqn (2.1). The analysis of the self-stress state in tensegrity prisms in Section 2.1 is such an example.

Matrix analysis

When the geometry is relatively complex, matrix analysis is necessary:

$$[G]\{F\} = \{0\} \tag{2.2}$$

where $[G]$ is the equilibrium matrix depending only on geometrical parameters, ranked $(3N_j - N_c) \times N_e$ for three-dimensional model, $\{F\}$ is the internal force vector, ranked $N_e \times 1$, and the right is the zero load vector applied to the unconstrained degrees of freedom, ranked $(3N_j - N_c) \times 1$. Therefore,

$$s = N_e - r_G \tag{2.3}$$
$$m = 3N_j - N_c - r_G \tag{2.4}$$

where r_G is the rank of equilibrium matrix $[G]$.

2.2.3 *Mechanisms in tensegrity simplexes*

All tensegrity prisms (truncated pyramids) are geometrically flexible. All infinitesimal mechanisms are first-order and thus initial stiffness can be obtained by one state of self-stress. In the triangular prism, $s = 1$, $m = 1$. The infinitesimal mechanism is a combination of rotation and translation between two bases and it is common for all tensegrity prisms and truncated pyramids. In the square prism, $s = 1$, $m = 3$, as it contains two additional infinitesimal mechanisms in two bases. In the pentagon, $s = 1$, $m = 5$, and so on. The characteristics can be proved by mechanism analysis.

Analysis of mechanical properties of pin-jointed systems including tensegrity prisms can be realized by the stiffness method introduced in the following section.

2.3 Analysis of pin-jointed systems containing mechanisms

In this section, the 'more conventional' stiffness-based Newton iteration method is used generally for analysing any pin-jointed systems in which components (bars and cables) are subjected to axial forces only (a cable component cannot remain straight under its own weight, but the effect is normally not critical), and (finite or infinitesimal) mechanisms are contained.

The analysis of cable networks contains two steps. One is the optional form-finding (also termed 'shape-finding') procedure aiming at determining the equilibrium geometry when self-stress is applied by shortening cables or lengthening struts. The other followed is the analysis of load response, the behaviour of the structure under load. Stiffness-based method itself (Argyris and Scharpf 1972) is not capable of analysing structures with finite mechanisms or high-order infinitesimal mechanisms, form-finding process when infinitesimal mechanisms exist, and unstable loading process when many cables slacken. These problems can be avoided by introducing dumb components that transform a structure with mechanisms to the equivalent 'geometrically rigid' structure. Therefore, the stiffness-based method can be all-powerful. Other analytical methods may be feasible, including force density method, dynamic relaxation method (Motro 1997), generalized inverse matrix method (Hangai and Lin 1989), linear complementary equation method (Wang *et al.* 1998), etc., but are out of the scope.

Main symbols to be used in the following section are listed as follows:

P	Point load at a joint
F	Internal force of an element
x, y, z	Nodal coordinates
l	Length of an element
E	Modulus of elasticity
A	Cross-sectional area of an element
R	Residual force at a joint.

2.3.1 *Formulation of equations*

The stiffness-based method can be explained by considering a node i in any spatial pin-jointed network connected to adjoining node j through elements e (Figure 2.4). The derivation is based on incremental loading process. Under load $P_{i,x}$, $P_{i,y}$, $P_{i,z}$ at the node i at the last step, the force in a typical element e is F_e and its length is l_e. The corresponding coordinates of nodes i and j are x_i, y_i, z_i and x_j, y_j, z_j, respectively.

At the initial state before external load is applied,

$$P_{i,x} = P_{i,y} = P_{i,z} = 0 \tag{2.5}$$

$$F_e = F_e^0 \tag{2.6}$$

in which, F_e^0 is the initial force under prestressed state (it equals to zero if prestress is not applied), which can be computed using the same method (see Section 2.3.2).

The equilibrium equations at node i at the previous loading step can be written as

$$\sum^e \frac{F_e}{l_e} \cdot (x_j - x_i) + P_{i,x} = 0 \tag{2.7a}$$

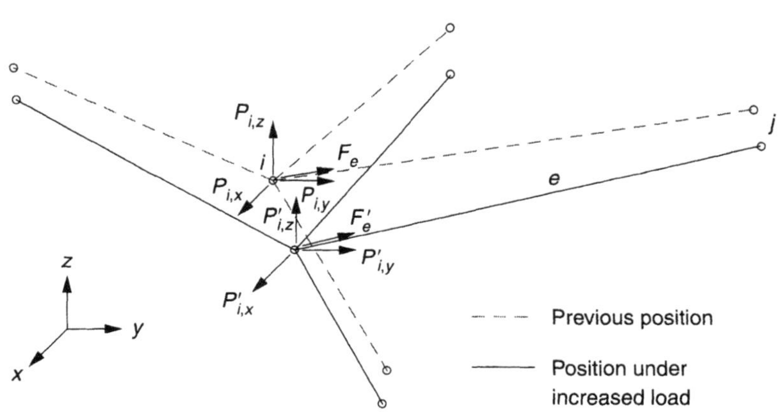

Figure 2.4 Force equilibrium diagram at an arbitrary joint i.

$$\sum^{e} \frac{F_e}{l_e} \cdot (y_j - y_i) + P_{i,y} = 0 \tag{2.7b}$$

$$\sum^{e} \frac{F_e}{l_e} \cdot (z_j - z_i) + P_{i,z} = 0 \tag{2.7c}$$

in which

$$l_e = \sqrt{(x_j - x_i)^2 + (y_j - y_i)^2 + (z_j - z_i)^2} \tag{2.8}$$

When the loads are increased to $P'_{i,x}$, $P'_{i,y}$, $P'_{i,z}$ by amounts $\delta P_{i,x}$, $\delta P_{i,y}$, $\delta P_{i,z}$, the nodes i and j deflect to x'_i, y'_i, z'_i through δx_i, δy_i, δz_i, and to x'_j, y'_j, z'_j through δx_j, δy_j, δz_j, respectively. Meanwhile, the force F_e and length l_e increase by δF_e to F'_e and δl_e to l'_e, respectively. Accordingly, Eqns (2.7) and (2.8) change as follows:

$$\sum^{e} \frac{F'_e}{l'_e} \cdot (x'_j - x'_i) + P'_{i,x} = 0 \tag{2.9a}$$

$$\sum^{e} \frac{F'_e}{l'_e} \cdot (y'_j - y'_i) + P'_{i,y} = 0 \tag{2.9b}$$

$$\sum^{e} \frac{F'_e}{l'_e} \cdot (z'_j - z'_i) + P'_{i,z} = 0 \tag{2.9c}$$

in which

$$l'_e = [(x_j + \delta x_j - x_i - \delta x_i)^2 + (y_j + \delta y_j - y_i - \delta y_i)^2 + (z_j + \delta z_j - z_i - \delta z_i)^2]^{1/2} \tag{2.10}$$

$$F'_e = F_e + \delta F_e = F_e + EA_e\left(\frac{l'_e}{l_e} - 1\right) \tag{2.11}$$

in which EA_e is the sectional modulus of element e. Neglecting the displacement items of higher order, we get

$$\frac{1}{l'_e} = \frac{1}{l_e}\left\{1 - \frac{1}{l_e^2}[(x_j - x_i)(\delta x_j - \delta x_i) + (y_j - y_i)(\delta y_j - \delta y_i) + (z_j - z_i)(\delta z_j - \delta z_i)]\right\} \tag{2.12}$$

Subtracting $\sum^e (F_e/l_e)(x_j - x_i)$ from both sides of Eqn (2.9a), $\sum^e (F_e/l_e)$ $(y_j - y_i)$ from both sides of Eqn (2.9b), and $\sum^e (F_e/l_e)(z_j - z_i)$ from both sides of Eqn (2.9c), and substituting Eqns (2.11) and (2.12) into Eqns (2.9a, b and c), respectively, we get

$$\left[\frac{F_e}{l_e}+\frac{EA_e-F_e}{l_e^3}(x_j-x_i)^2\right](\delta x_j-\delta x_i)+\left[\frac{EA_e-F_e}{l_e^3}(x_j-x_i)(y_j-y_i)\right](\delta y_j-\delta y_i)$$

$$+\left[\frac{EA_e-F_e}{l_e^3}(x_j-x_i)(z_j-z_i)\right](\delta z_j-\delta z_i)=-R_{i,x} \qquad (2.13a)$$

$$\left[\frac{EA_e-F_e}{l_e^3}(y_j-y_i)(x_j-x_i)\right](\delta x_j-\delta x_i)+\left[\frac{F_e}{l_e}+\frac{EA_e-F_e}{l_e^3}(y_j-y_i)^2\right](\delta y_j-\delta y_i)$$

$$+\left[\frac{EA_e-F_e}{l_e^3}(y_j-y_i)(z_j-z_i)\right](\delta z_j-\delta z_i)=-R_{i,y} \qquad (2.13b)$$

$$\left[\frac{EA_e-F_e}{l_e^3}(z_j-z_i)(x_j-x_i)\right](\delta x_j-\delta x_i)+\left[\frac{EA_e-F_e}{l_e^3}(z_j-z_i)(y_j-y_i)\right](\delta y_j-\delta y_i)$$

$$+\left[\frac{F_e}{l_e}+\frac{EA_e-F_e}{l_e^3}(z_j-z_i)^2\right](\delta z_j-\delta z_i)=-R_{i,z} \qquad (2.13c)$$

in which nodal residual in-equilibrium force $R_{i,x}$, $R_{i,y}$ and $R_{i,z}$ is given as follows:

$$R_{i,x}=P'_{i,x}+\sum^{e}\frac{F_e}{l_e}(x_j-x_i) \qquad (2.14a)$$

$$R_{i,y}=P'_{i,y}+\sum^{e}\frac{F_e}{l_e}(y_j-y_i) \qquad (2.14b)$$

$$R_{i,z}=P'_{i,z}+\sum^{e}\frac{F_e}{l_e}(z_j-z_i) \qquad (2.14c)$$

Applying Eqn (2.13) to all joints, we get

$$[K]\{d\}=\{R\} \qquad (2.15)$$

in which $[K]$ is the stiffness matrix ranked $3N_j \times 3N_j$ (N_j is the total number of joints), $\{d\}$ is the displacement vector ranked $3N_j \times 1$, and $\{R\}$ is the nodal in-equilibrium force vector ranked $3N_j \times 1$. $[K]$ should always be non-singular.

In Eqn (2.13), F_e is normally negligible compared to EA_e for most materials (e.g. the proportion is less than 1% for steel). The items containing EA_e (e.g. $((EA_e-F_e)/(l_e^3))(x_j-x_i)^2$, etc.) form the elastic stiffness, and the item F_e/l_e forms the geometrical stiffness. When a system is loaded at its infinitesimal mechanisms, only geometrical stiffness due to prestress contributes before elastic stiffness due to deformation is developed. So the stiffness of the system with infinitesimal mechanisms is low, as proved in Section 2.4.1.

2.3.2 Computation process

For a geometrically flexible system, incremental loading process is generally required for the stability of the computation process. The iteration process for a loading step is given as follows:

(a) Form $\{R\}$.
(b) Form $[K]$.
(c) Solve Eqn (2.15) to get $\{d\}$.
(d) Get new coordinates and internal forces.
(e) Form $\{R\}$. If it does not converge and the iteration has not reached the prescribed circles, return to b.

Form-finding process under prestress $\{F^0\}$, if introduced, can be calculated based on the same method with zero load vector.

2.3.3 Singularity in stiffness matrix and dumb element method

Singularity in stiffness matrix

Problems in using the stiffness method happen when $[K]$ cannot be non-singular. It happens at the following cases:

- *Form-finding process for geometrically flexible systems with first-order mechanisms.* In a system of first-order infinitesimal mechanisms, the initial stiffness (geometrical stiffness) can be provided by a state of self-stress, after that $[K]$ becomes non-singular. When the geometry is simple, the resulting internal forces under the self-stress state can be input directly, and the resulting deformation is neglected (e.g. the cable string in Figure 2.5). When the geometry is complex, how to obtain the resulting internal forces and geometry becomes a problem. One aspect of the problem is that $[K]$ is singular before the prestress is introduced. The other is that during the self-stressing process, bars or cables for introducing prestress actually do not contribute stiffness to the system. But if these components are removed in the computation model and replaced by prestress forces, new mechanisms are produced (even an originally geometrically rigid system may be changed into a geometrically flexible/deformable system). An existing method is to keep these components in the iteration process (Hanaor 1992). However, the process is hard to converge.
- *A system with high-order infinitesimal mechanisms or finite mechanisms.* A system with high-order infinitesimal mechanisms cannot be rigidified by one state of self-stress. A system with finite mechanisms cannot be rigidified at all. However, these mechanisms may actually reach a stable final position under load after experiencing large displacement (including rigid body movement) despite that singularity in stiffness matrix hinders efficient analysis.

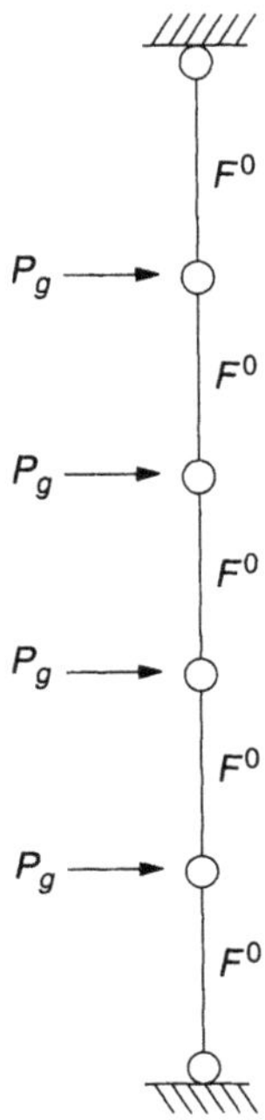

Figure 2.5 A cable string (containing first-order infinitesimal mechanisms).

- *Loading process due to cable slackening.* During the loading process, some cables may slacken. They cannot sustain forces and thus behave as they are temporarily removed. The number and position of slack cables vary with load. The stiffness matrix may become singular (calculation process becomes suddenly unstable) after these cables are 'removed' through affording zero stiffness and forces, even if the initial geometry is geometrically rigid. At the state, it is hard to say whether the structure is actually stable or not.

Dumb component method

The analysis of the geometrically flexible/deformable form can be realized by analysing its equivalent geometrically rigid form, which is formed by introducing the additional components, namely, dumb components of small stiffness and zero stress (Figure 2.6). The internal forces of dumb components are always restored to zero for iteration.

In actual structural design, the length of the dumb component is set the same degree of magnitude as the existing elements in the structure. As for the cross-sectional area, if it is too small, the computation process may be unstable. But if not sufficiently small, accuracy requirement cannot be satisfied and the iteration process may not converge. The cross-sectional area

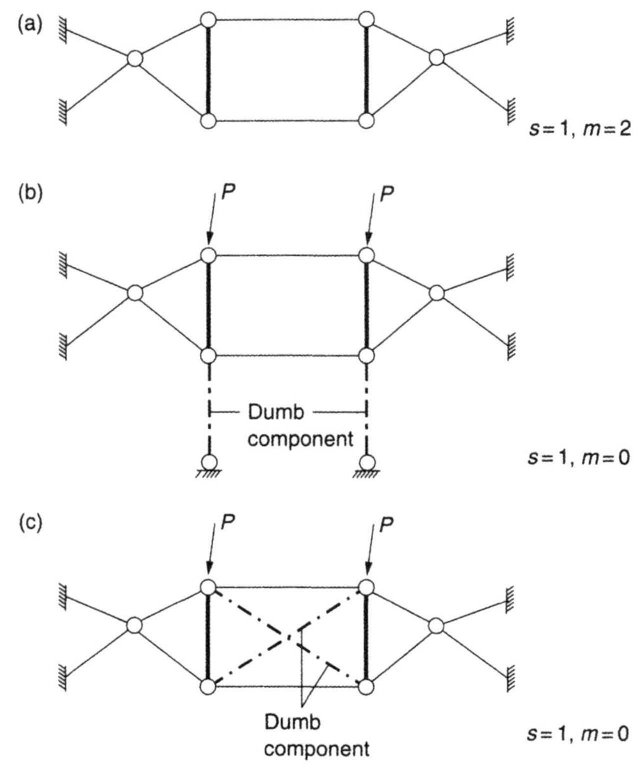

Figure 2.6 Concept of introducing dumb components: (a) geometrically flexible; (b) and (c) equivalent form to (a).

recommended is 0.1~1% that of the smallest element in the structure. The accuracy is in general almost 100%.

When dumb components are introduced, one joint of a dumb component is attached to the joint where (infinitesimal) mechanisms possibly exist. The other joint may be supported (Figure 2.6(b)) or connected to an existing node (Figure 2.6(c)). If we are not sure exactly how many mechanisms and where these mechanisms exist, it does not matter introducing more dumb components. For example, as the extreme case we may introduce three orthogonal dumb components at each joint. If so, it can be automatically realized by a program without the need to identify the mechanisms. Despite that the method increases joints and elements, it is not a problem for modern computers and actually such system is normally not of complex geometry.

2.3.4 *Numerical examples*

Systems containing various kinds of mechanisms can be transformed into the equivalent geometrically rigid form by introducing dumb components.

In addition to the following cases, the idea can also be extended to analyse deployable structures and membrane structures.

Form-finding for structures with infinitesimal mechanisms

In Figure 2.7, is presented a regular square-based tensegrity prism. The supports provided are only to avoid rigid body movements as a whole. The prism contains a state of self-stress ($s = 1$) and three states of infinitesimal mechanisms ($m = 3$). Self-stress is supposed to be introduced by elongating struts to reach 100kN. For all struts, sectional stiffness is 2×10^6kN; for all cables, 1.8×10^5kN. As the initial stiffness matrix is singular, form-finding process cannot proceed.

To avoid the problem, four dumb components are introduced to the original geometry (Figure 2.7(a)). Their sectional stiffness is 1% that of cables, 1.8×10^3kN. In addition, four struts are also afforded small stiffness to let the computation process converge quickly. The value is one-twentieth of the original, 1×10^5kN. The resulting internal forces are accurate as given in the resulting geometry (Figure 2.7(b)).

Systems with finite or high-order infinitesimal mechanisms

In Figure 2.8, it is presented an anchored string of bars with finite mechanisms. At the beginning, the components are in a 'disordered' state (Figure 2.8(a)). When load is introduced, they will move to the equilibrium state. Based on the present method, three dumb components are introduced to the

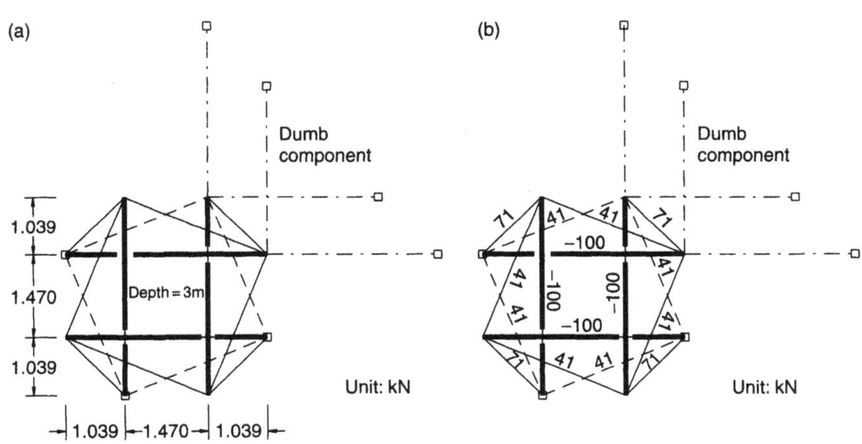

Figure 2.7 Form-finding of a tensegrity prism by dumb component method: (a) a square-based tensegrity prism; (b) internal forces under 100kN to struts.

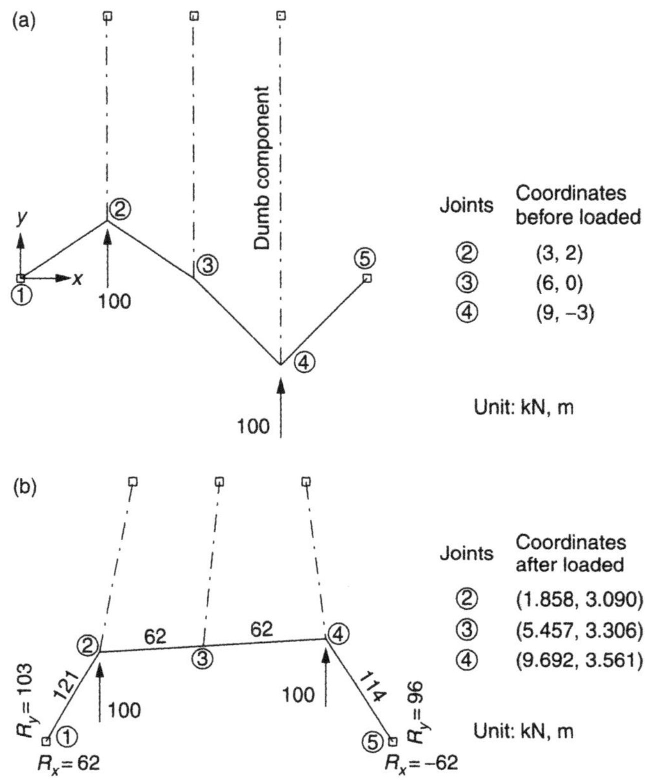

Figure 2.8 The analysis of a geometrically deformable system by dumb component method: (a) original geometry and load case; (b) geometry and internal forces under load.

original geometry to transform the model into a geometrically rigid one. The sectional stiffness for all bars is 2×10^6kN; for the dumb components, 2×10^3kN. The resulting equilibrated internal forces under final geometry are given (Figure 2.8(b)).

The method is the same for calculating the load response of structures with high-order infinitesimal mechanisms such as the example in Figure 2.9. The transformed form is given whereas the analysis is omitted here.

Loading processes

Following the principle of applying dumb components, slack cables in all cable structures are afforded a small stiffness. Therefore, computation process is stable if the actual structure is stable. Examples are omitted.

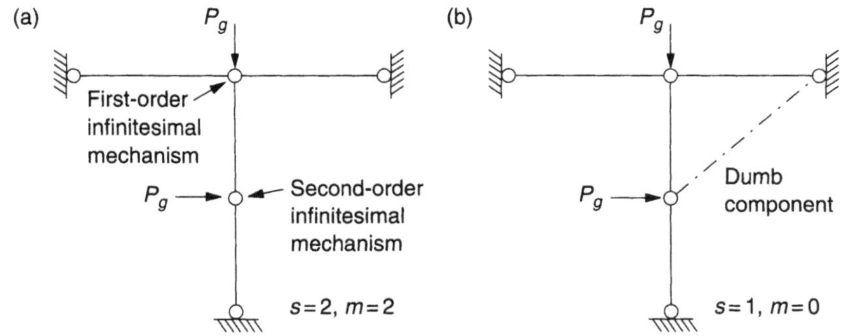

Figure 2.9 Dumb component method in high-order infinitesimal mechanisms: (a) high-order geometrically flexible system; (b) transformed geometrically rigid system.

2.4 Mechanical properties related to tensegrity simplexes

Tensegrity prisms are geometrically flexible. General properties of geometrically flexible systems are discussed in this section, including how to transform these systems into geometrically rigid ones. Following that, rigidified tensegrity simplexes are obtained.

2.4.1 Properties of geometrically flexible systems

The general properties of geometrically flexible systems, including tensegrity prisms, can be explained by the simplest planar two-cable model in Figure 2.3(c) or Figure 2.10(a) ($s = 1, m = 1$). The load applied can be divided into two components: the elastic load and the geometrical load. The elastic load (P_e), also termed extensional load or fitted load, produces only elastic deformations without activating the mechanisms – it is fitted to the initial geometry. The geometric load (P_g), also termed inextensional load, causes changes in the geometry through activation of the mechanism but no elastic strains (based on the initial geometry).

Given $P_e = 0$, $P_g = 0.0001$EA, different levels of prestress (F_p) are introduced into cables in order to balance load under deformation (when $F_p = 0$, dumb component method is applied). The internal forces (F) and displacements (d) obtained are presented in Figure 2.10(b). It shows that a geometrically flexible system produces large internal forces under geometrical load and high prestress is required to control displacement. The load–displacement curve under the non-prestress condition ($F_p = 0$) is shown in Figure 2.11(c) (the $D/L = 0$ case). It presents zero stiffness at the start under non-prestress

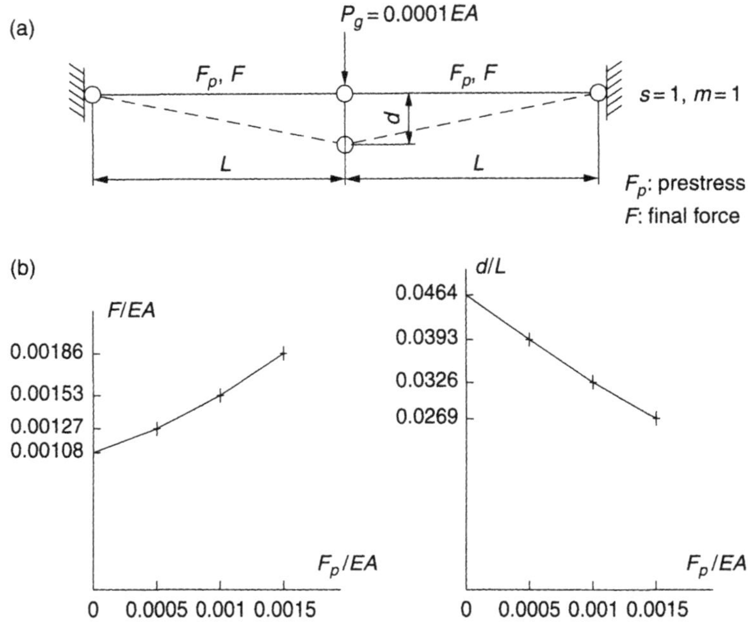

Figure 2.10 Properties of planar two-cable model with mechanisms: (a) planar two-cable model with infinitesimal mechanism; (b) internal force and deflection under different prestress levels.

condition. Then, the stiffness keeps increasing, showing 'strain-hardening' phenomenon, as the geometrical stiffness component is dominant and increases with internal forces. After experiencing large deformation (the deformed geometry is already 'away' from mechanism), the stiffness becomes stable as the elastic stiffness component becomes dominant.

The infinitesimal mechanism can be considered as the limit case of *near-mechanism geometry*. It is presented for reference in Figure 2.11 the internal forces and displacement of the near-mechanism geometry (when sag D is approximate to zero) compared with those of 'normal' geometry (when D is not very small). Near-mechanism geometry is also characteristic of large internal forces and especially large displacement compared with 'normal' geometry, indicating insufficient utilization of material strength. The effect of strain hardening gradually disappears when the geometry becomes 'farther' from mechanism (here, strain hardening becomes negligible when D/L = 0.1, see Figure 2.11(c)). Near-mechanism case often happens in conventional cable networks, whose lightness is mainly due to high strength of cables and boundary anchoring system (to which the load is transferred).

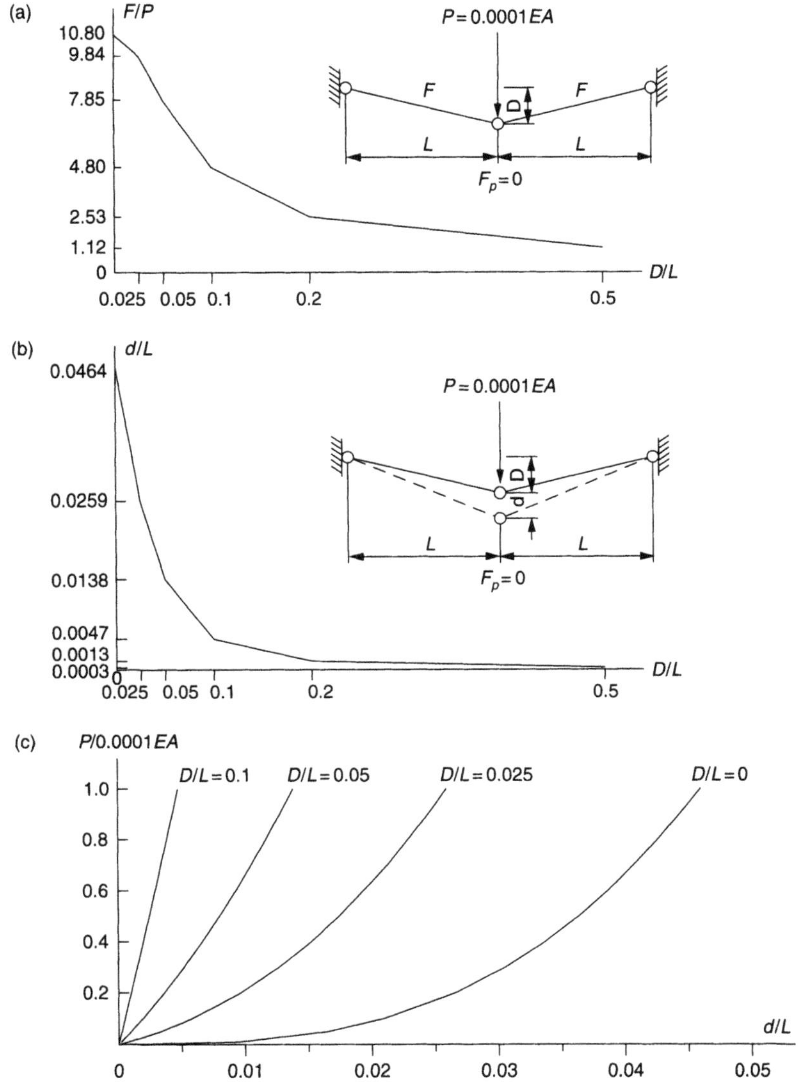

Figure 2.11 Mechanical properties of two-cable model with different sags: (a) internal force; (b) deflection; (c) load–displacement curve.

2.4.2 *Principles of transforming geometrically flexible systems to rigid systems*

A geometrically flexible bar system can be transformed into a geometrically rigid one by adding components and/or changing the geometry to remove

Cases	Equivalent two-dimensional case (two-cable model)	Triangular prism
(a) Geometrically flexible	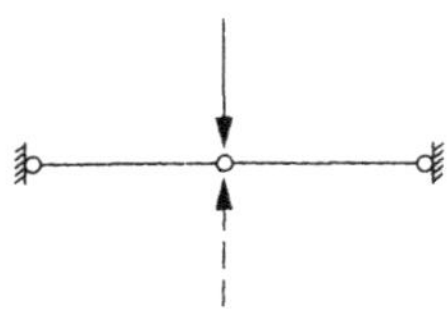	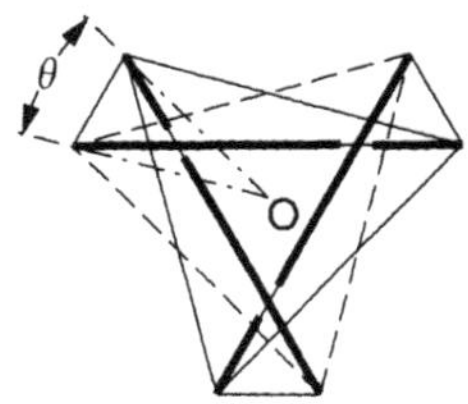 $\theta = 30°$
(b) Changing geometry only (unstable)	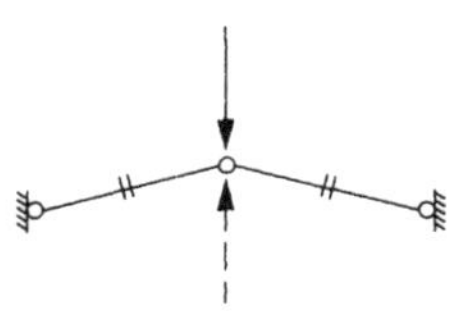	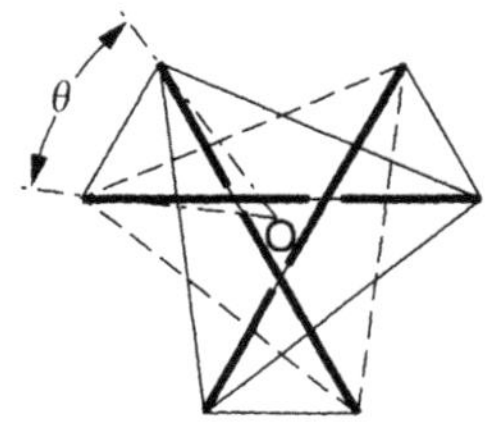 $30° < \theta < 60°$
(c) Adding cables only (poor geometrically rigid case)	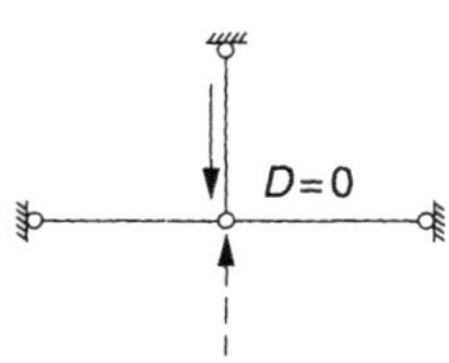 $D=0$	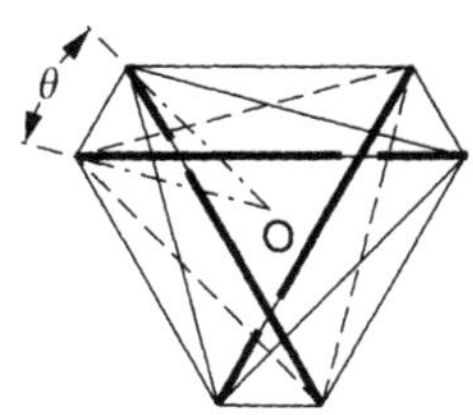 $\theta = 30°$
(d) Geometrically rigid (adding cables and changing geometry)	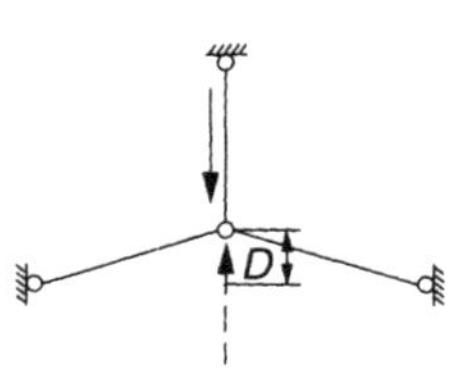 $0 < D$	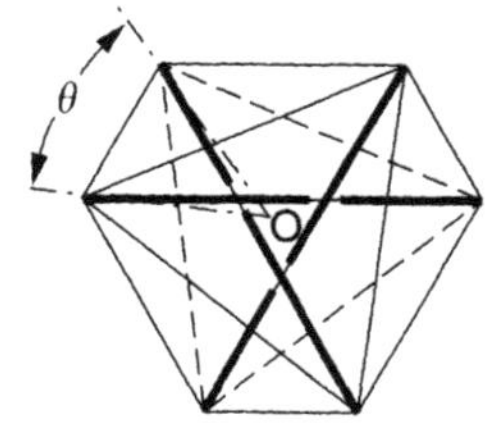 $30° < \theta < 60°$

Figure 2.12 Forming rigidified tensegrity simplexes.

the mechanisms. In cable networks, the case is relatively complicated as cables cannot sustain compression and thus may become slack.

In a geometrically flexible cable network, changing the geometry can let the system superficially satisfy geometrically rigid condition. However, cables may slacken at the deformed geometry, producing finite mechanisms (Figure 2.12, case (b)). So the method is normally not applicable.

Adding cables without changing the geometry can give the system initial stiffness as can be expressed in the stiffness matrix, satisfying the geometrically rigid condition. The added cable may slacken under certain loads and thus the original mechanism still '*implicitly*' exists (Figure 2.12, case (c)). But anyway, adding cables reduces the chance of activating the mechanism. Generally, it is 'safe' both to change the geometry and to add components in order to stabilize the deformed geometry (Figure 2.12, case (d)).

A geometrically rigid system unavoidably produces slack cables under load. It is generally required that the geometrically rigid system be stable under applied loads after cable slackening. The number and position of slack cables vary with loads and there is no generalized rule to evaluate it. Prestress can postpone the occurrence of cable slackening but cannot ultimately avoid it unless the level is excessively high compared with the external load. When prestress level is low, the effect on internal forces and thus component design is generally not significant for a geometrically rigid system.

2.4.3 Rigidified tensegrity simplexes

Based on the earlier discussion, rigidifying tensegrity simplexes necessitates the introduction of lateral diagonal cables. In the triangular prism of increased prism rotation angle (Figure 2.12, case (d)), adding one lateral cable can superficially satisfy the geometrically rigid condition. However, as

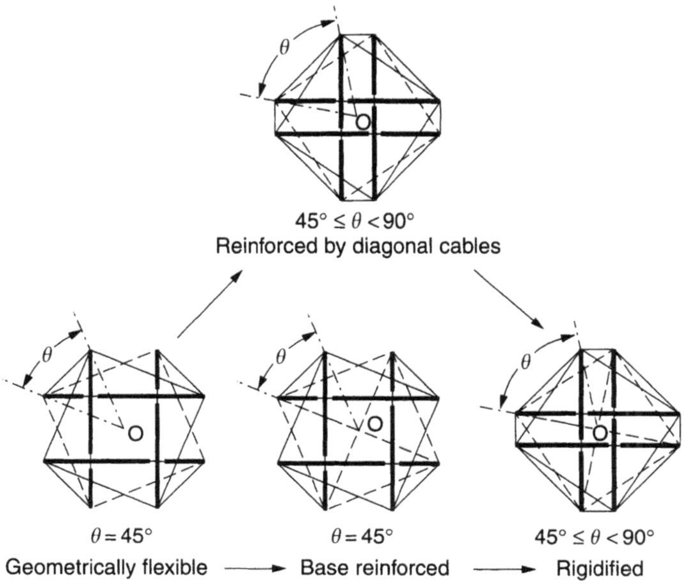

Figure 2.13 Forming rigidified square tensegrity simplex.

instability occurs in the left joints owing to potential cable slackening, three lateral cables are added in the resulting rigidified geometry, which contains three states of self-stress based on the initial geometry.

The introduction of lateral additional cables removes the infinitesimal mechanism of prism rotation, but the infinitesimal mechanisms in the base polygon higher than the triangular cannot be avoided. In the rigidified square prism (Figure 2.13), for example, in addition to four additional lateral diagonal cables, two crossing cables are introduced in one base to remove two mechanisms in the bases. It contains four states of self-stress.

In the tensegrity prisms reinforced by lateral diagonal cables (for the triangular, rigidified), the relative rotation angle of the two bases is not unique and can be varied between the original prism rotation angle, at which the additional cables cannot be self-stressed (geometry unchanged), and at the limit state, double that angle, at which the struts would intersect at the centre (for the triangular: 30° to less than 60°; for the square: 45° to less than 90°). With the increase of the prism rotation angle, lateral additional cables are apt to share more forces although the actual behaviour depends on how external forces are applied.

3 Structural configurations, properties and design of tensegrity grids composed of simplexes

Fuller believed that tensegrity is 'the rule of nature'. He thought that tensegrity structures are naturally optimal structural forms without testifying it. His belief has significant influence on his followers. Morphological studies widely spread out to express ultimately the impact of tensegrity concept on architecture. But in contrast, application studies on structural behaviour were carried out only in recent years. Among tensegrity forms, double-layer tensegrity grids composed of tensegrity simplexes appear most suitable as structural forms and are illustrated in detail in this chapter, including configuration details related to the geometrical characteristics, design examples presenting mechanical properties, and evaluating tensegrity concept by the law of mechanics. The analysis shows the inherent deficiency of tensegrity structures. Accordingly, principles on how to achieve structurally efficient systems are summarized. Architectural aspect of tensegrity systems regarding other tensegrity forms is presented in Section 7.1.

3.1 Structural systems made of tensegrity simplexes

As structural building blocks, tensegrity simplexes can be used in various planar layouts. The configurations made of higher simplexes (especially geometrically rigid forms) induce complex geometry and thus are not so convenient. Therefore, triangular and square simplexes are studied representatively. In general, triangular simplexes are suitable for circular and elliptical plans, whereas square simplexes are suitable for square and rectangular plans. Structural configurations composed of tensegrity simplexes include non-contiguous strut configurations and contiguous strut ones, as illustrated as follows. In these double-layer forms, struts are sandwiched between two parallel (when flat) cable surfaces as skeletons to brace the volume.

3.1.1 Non-contiguous strut configurations

Non-contiguous strut configurations (or non-bar-to-bar connections) are different from 'regular' connection methods that simplexes are connected

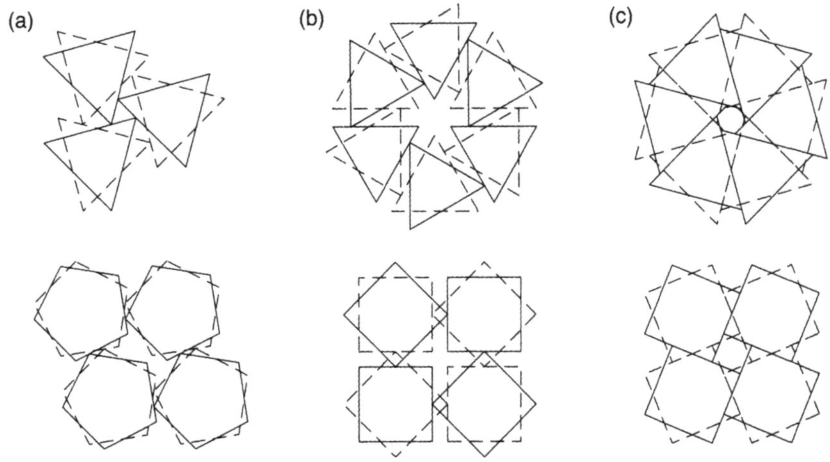

Figure 3.1 Connecting methods for non-contiguous strut configurations by Hanaor: (a) vertex-to-edge connection (Method Ia); (b) vertex-to-edge connection (Method Ib); (c) edge-to-edge connection (Method II).

directly at vertex joints. On the contrary, tensegrity simplexes are connected in a node-to-cable way. Therefore, each connected cable is divided into two and struts are not in contact. The resulting non-contiguous strut configurations do not contain continuous struts and thus are 'pure' tensegrity forms. Emmerich (1990) was the first to conceive non-contiguous strut tensegrity grids. Hanaor developed the concept and proposed the way of connecting simplexes into three methods (Ia, Ib and II, see Figure 3.1). Methods Ia and Ib are vertex-to-edge connections of prisms. In Method Ia, prisms are connected consecutively (Figure 3.1(a)). Between each pair of adjacent tensegrity prisms, one prism is linked to another by vertex-to-edge connection in both bases. The method is feasible only for prisms with odd number of vertices in the base polygon. In Method Ib, each pair of prisms is connected by linking in one layer the vertex of one prism to the edge of the other prism and alternatively in the other layer (Figure 3.1(b)). The last (Method II) is the edge-to-edge connection, in which adjacent prisms share a segment of the edge in each layer (Figure 3.1(c)).

In the configurations made of triangular tensegrity prisms, Method Ia has better stiffness than Methods Ib and II and the infinitesimal mechanism between two bases is avoided by the restraints from six adjacent simplexes. It has been studied systematically by Hanaor (1991a,b, 1992, 1994, 1997). In domical forms, tensegrity truncated pyramids are employed as building blocks in place of tensegrity prisms (Figure 3.2).

In the configurations made of square tensegrity prisms, Method Ia is not applicable. In Method Ib or II, unlike in the triangular prism, struts (and

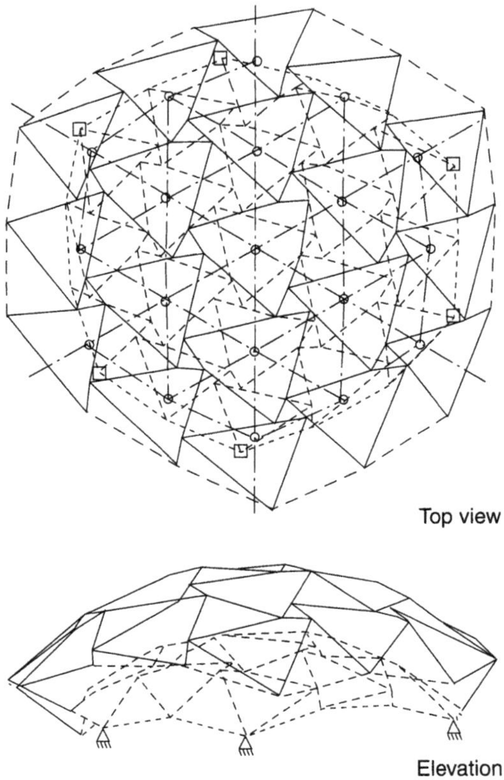

Figure 3.2 Double-layer tensegrity dome of Method Ia.
Source: Hanaor (1992); Courtesy: Multi-Science Publishing.

also diagonal cables) in each square prism can be placed in various ways due to 'chirality' or 'handedness', presenting various architectural forms. Method Ib produces two layouts: regular layout (Figure 3.3(a)) and irregular layout (Figure 3.3(b)). In the regular layout, simplexes are copied consecutively. In Method II, three layouts are identified including Layout A – strut orthogonal to the edge (Figure 3.4(a)), Layout B – strut diagonal to the edge (Figure 3.4(b)), and Layout C – mixed type (Figure 3.4(c)). Cable interference may occur in some layouts. In the layout in Figure 3.4(c), unlike in other layouts, prism rotation angle greater than 45° is not feasible.

3.1.2 Contiguous strut configurations

The other way to form tensegrity grids, which was proposed by Motro, is node-to-node connection of simplexes (Motro 1990). Therefore, struts are

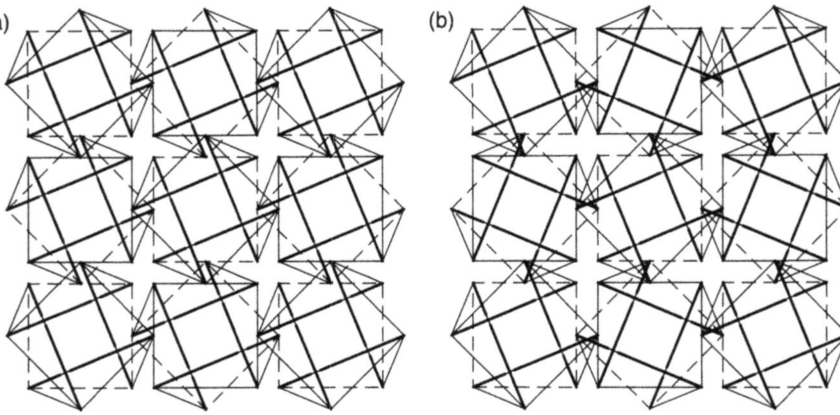

Figure 3.3 Layouts of non-contiguous strut configuration Method Ib for the square simplex: (a) regular layout (geometrically flexible); (b) irregular layout (geometrically flexible).

Source: Wang and Li (2001); Courtesy: Multi-Science Publishing.

connected directly at ends. Contiguous strut configurations (or bar-to-bar connections) violate 'islands of compression in a sea of tension' of the original tensegrity concept but they are included within tensegrity systems as the extended concept on the premise that the configurations are based on tensegrity simplexes.

There are basically two methods to form contiguous strut configurations for low tensegrity simplexes. One is the vertex-to-vertex (two-vertex) connection (Method I, Figure 3.5(a)) (Wang and Liu 1998). Simplexes are connected at vertices to each other in both layers with no common edges. The other, as recommended by Motro, is the vertex-and-edge (three-vertex) connection of simplexes (Method II, Figure 3.5(b)). Each pair of adjacent simplexes shares a vertex in the layer with small base and an edge (two adjacent vertices) in the other layer. In the flat form made of triangular or square simplexes, tensegrity prisms are suitable for the vertex-to-vertex connection, and tensegrity truncated pyramids suitable for the vertex-and-edge connection. The configurations of the pentagonal and hexagonal prisms are presented (mainly as architectural forms) in Figure 3.5(c)–(e) for reference. It is interesting to note that the 'prism' form of the hexagonal tensegrity prisms is connected in the face-to-face way (Figure 3.5(a)).

In the vertex-and-edge connection (Method II) of square tensegrity truncated pyramids, two feasible layouts of struts can be presented due to the chirality. One is the regular layout (Layout A) in which struts are connected co-linearly in plan (Figure 3.6(a)), the other is the irregular layout (Layout B) in which struts are connected in a zig-zag way in plan (Figure 3.6(b)).

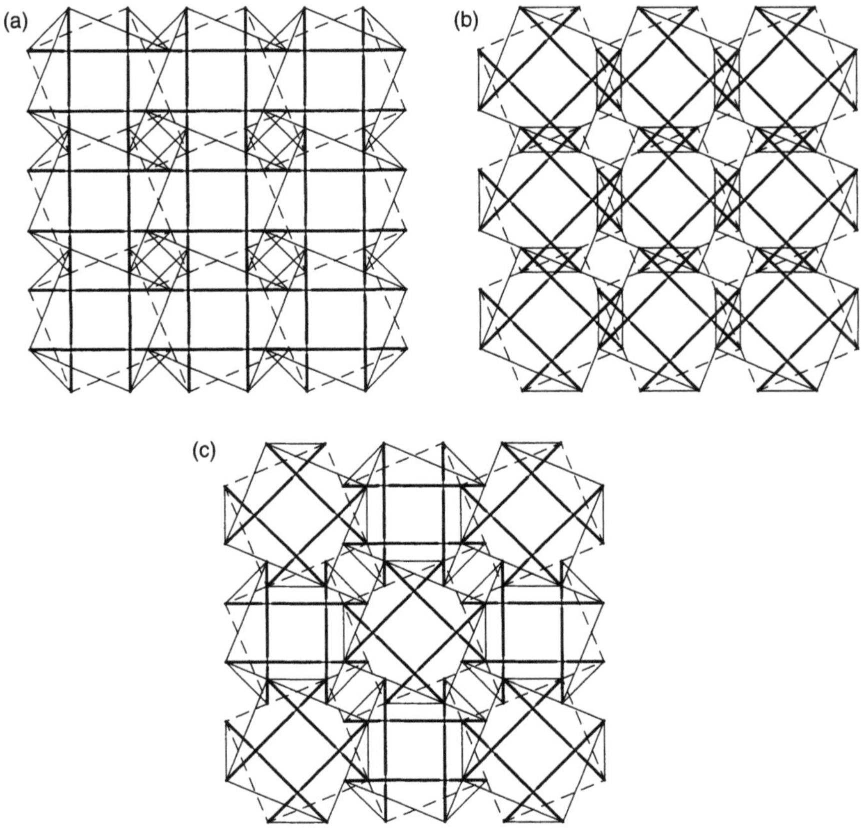

Figure 3.4 Layouts of non-contiguous strut configuration Method II for the square simplex: (a) struts orthogonal to the edge (Layout A, geometrically flexible); (b) struts diagonal to the edge (Layout B, geometrically flexible); (c) mixed type (Layout C, geometrically flexible).

Source: Wang and Li (2001); Courtesy: Multi-Science Publishing.

In the vertex-to-vertex connection (Method I) of square tensegrity prisms, struts can be introduced typically in three directions due to the chirality. In Layout A, struts are orthogonal to the edge (Figure 3.7(a)), in Layout B, struts being diagonal to the edge (Figure 3.7(b)). Layout C is the mixed type in which only 45° prism rotation angle is feasible (Figure 3.7(c)).

3.2 Geometrically rigid forms

Geometrically rigid forms have better distribution of internal forces and thus avoid large values. This is beneficial as joint design and standardization of fabrication become easier. What is more important, deflection is

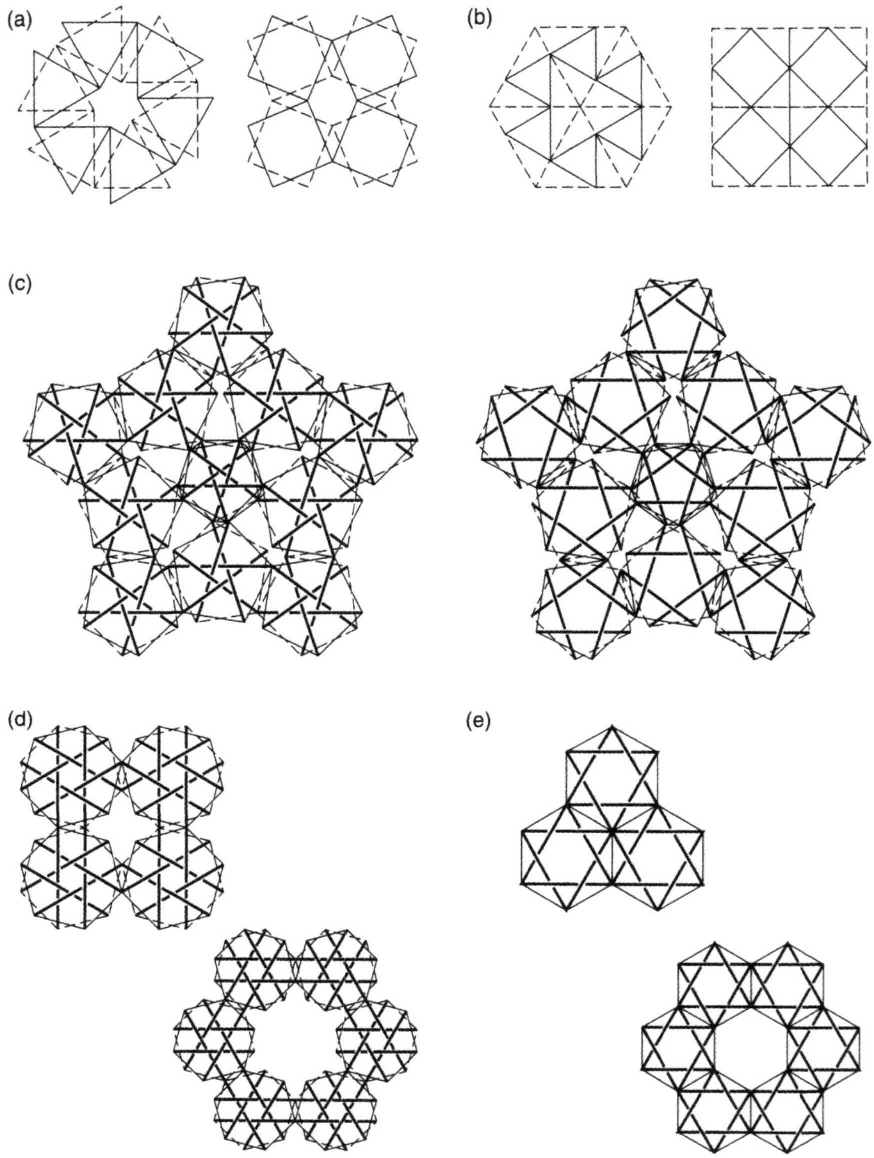

Figure 3.5 Connecting methods for contiguous strut configurations: (a) vertex-to-vertex connection (Method I); (b) vertex-and-edge connection (Method II); (c) vertex-to-vertex connection of pentagonal tensegrity prism (a small simplex at the centre); (d) vertex-to-vertex connection of hexagonal tensegrity prism; (e) face-and-face connection of hexagonal tensegrity prism.

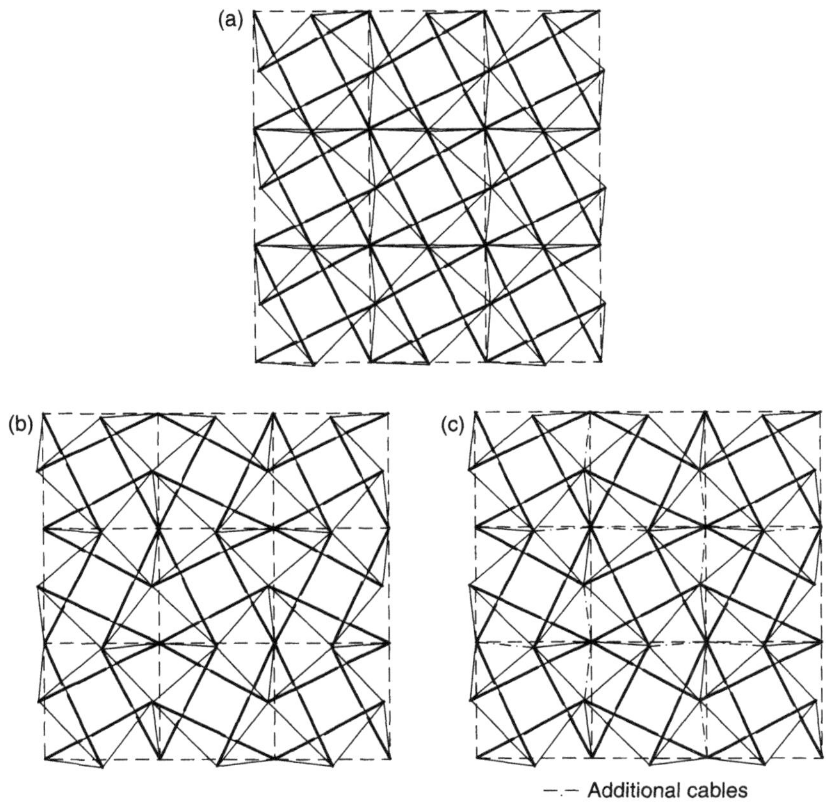

Figure 3.6 Layouts of contiguous strut configuration Method II for the square truncated pyramid: (a) regular layout (Layout A, geometrically rigid form); (b) irregular layout (Layout B, geometrically flexible form); (c) irregular layout (Layout B, geometrically rigid form).

Source: Wang and Li (2001); Courtsey: Multi-Science Publishing.

much smaller (e.g. reduced often by about half in non-contiguous strut grids). The only drawback, which is unavoidable, is that more cables are added. However, the gross weight is still reduced based on the much improved load-transfer pattern. It is recommended that in 'non-structural' application we use geometrically flexible forms, whereas in 'structural' application, we transform them into geometrically rigid forms. Therefore, prestress is not indispensable for tensegrity structures.

3.2.1 Non-contiguous strut configurations

In non-contiguous strut configurations, dividing each cable in the base into two when connecting adjacent simplexes results in two infinitesimal mechanisms.

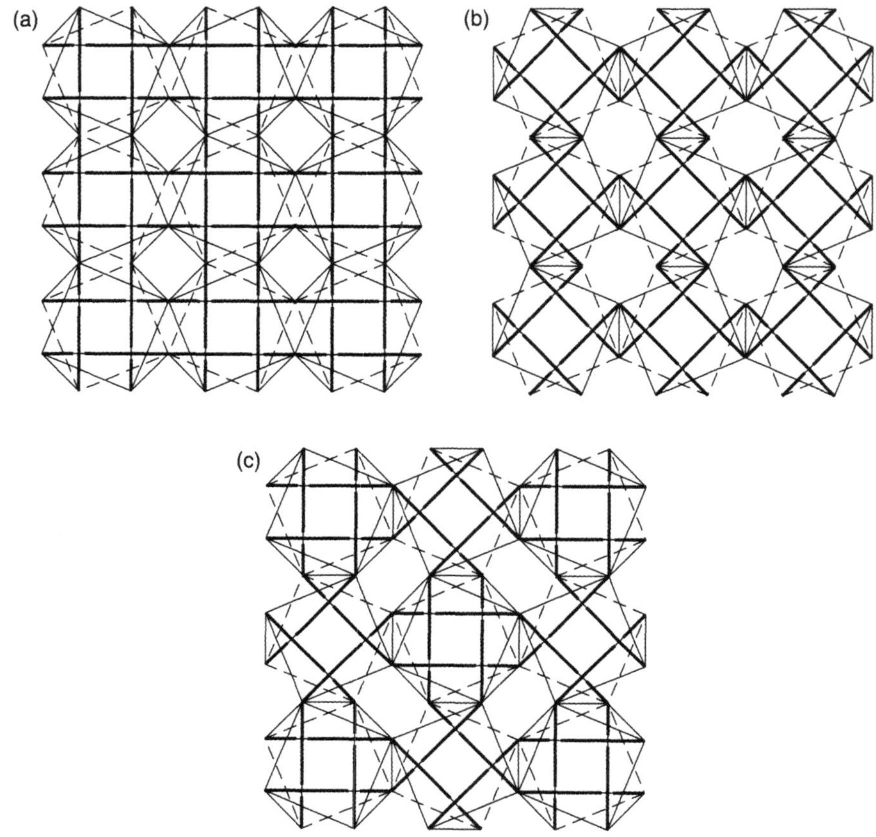

Figure 3.7 Layouts of contiguous strut configuration Method I for the square prism: (a) Layout A – struts orthogonal to the edge (geomatrically flexible); (b) Layout B – struts diagonal to the edge (geometrically flexible); (c) Layout C – mixed type (geometrically flexible).

Source: Wang and Li (2001); Courtesy: Multi-Science Publishing.

Such mechanisms cannot provide initial restraints to the connected vertex. So even the grid formed by joining geometrically rigid prisms is still geometrically flexible although it is stiffer (Figure 3.8(b)). These infinitesimal mechanisms, as proposed by Hanaor (1994), can be avoided by attaching cables in different layers to adjacent simplexes.

In the geometrically rigid configurations (Method Ia) made of geometrically flexible triangular tensegrity prisms (Figure 3.8(c)), each connecting vertex receives two diagonal attaching cables (each cable is attached to two layers). Some inclined cables can be appended at the boundary to reduce maximum strut forces. In the geometrically rigid grid, increased

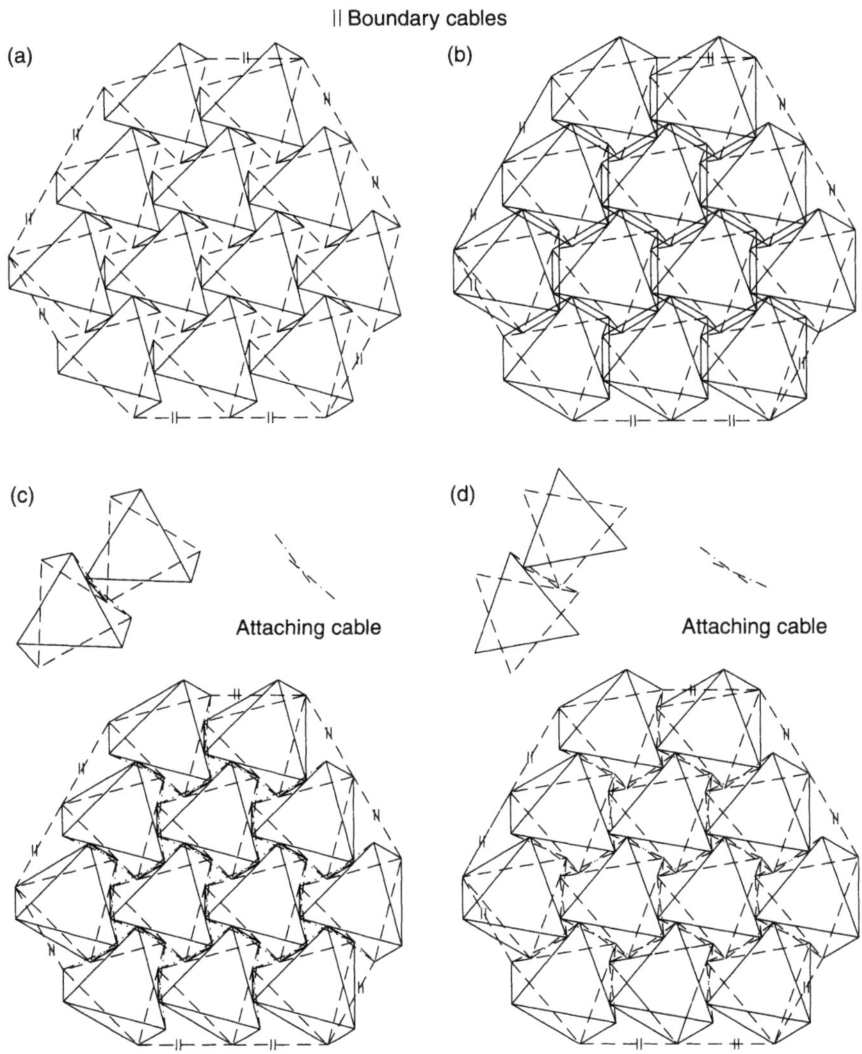

Figure 3.8 Layouts of non-contiguous strut configuration Method Ia for the triangular simplex: (a) geometrically flexible form composed of geometrically flexible simplexes; (b) geometrically flexible form composed of geometrically rigid simplexes; (c) geometrically rigid form composed of geometrically flexible simplexes; (d) geometrically rigid form composed of geometrically rigid simplexes.

Source: Wang and Li (2001); Courtesy: Multi-Science Publishing.

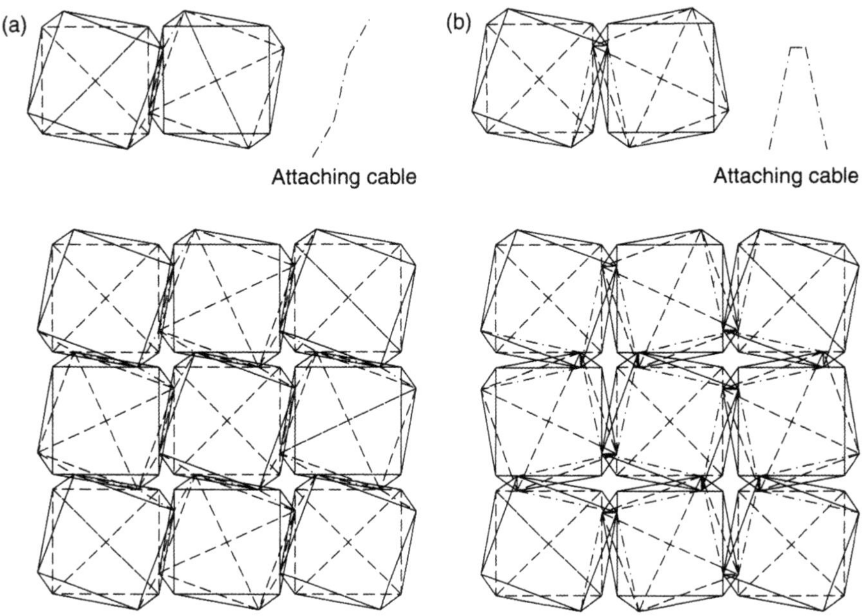

Figure 3.9 Geometrically rigid layouts of non-contiguous strut configuration Method Ib for the square simplex: (a) regular layout (geometrically rigid); (b) irregular layout (geometrically rigid).

Source: Wang and Li (2001); Courtesy: Multi-Science Publishing.

prism rotation angle can be designed with better mechanical properties and all inclined cables except those at the boundary are removable as their forces can be transferred to attaching cables. Hence, Hanaor's geometrically rigid form is derived. The grid with 40° prism rotation is given in Figure 3.8(d).

In the grid composed of square tensegrity prisms, each inner simplex has only four adjacent simplexes to restrain it compared with six in the grid composed of triangular tensegrity prisms (Method Ia), thus the infinitesimal mechanism between two bases cannot be avoided. In addition, geometrically flexible square tensegrity prisms contain mechanisms in the bases, so geometrically rigid prisms have to be employed to form geometrically rigid grids. The new mechanisms owing to prism connection can be removed by the same principle – attaching diagonal cables to different prisms. The resulting geometrically rigid forms are given in Figures 3.9(a), (b) and 3.10, in which prism rotation angles are 70° for Method Ib and 45° for Method II, respectively. Some inclined cables always slacken and can be removed. But the case is complex and varies with layouts, so the topic is out of the scope.

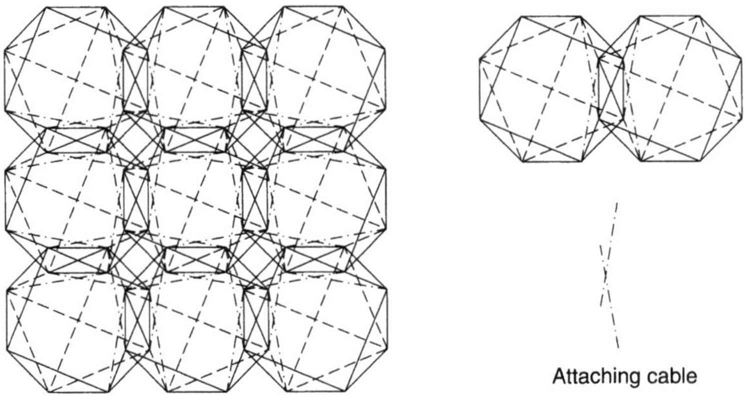

Figure 3.10 Geometrically rigid form for all layouts of non-contiguous strut configuration Method II for the square simplex.

Source: Wang and Li (2001); Courtesy: Multi-Science Publishing.

3.2.2 Contiguous strut configurations

In the vertex-and-edge connection of the square tensegrity truncated pyramid, mechanisms in the bases are avoided by the restraints from adjacent simplexes. In the irregular layout, additional diagonal cables are required to compensate for shared cables so as to remove mechanisms (Figure 3.6(c)). The regular layout is geometrically rigid (when the number of modules is at least three along each way) since no diagonal cables are shared by adjacent simplexes, but it does not contain higher prism rotation angle.

The vertex-to-vertex connection produces more joints and possibility of mechanisms. In the connection of the square tensegrity prism, the mechanisms in simplexes cannot be removed and the resulting grid is geometrically flexible. In order to form geometrically rigid grids, rigidified simplexes have to be employed although some cables may be taken away (Figure 3.11). In addition, it is interesting to note that in the grid there is still one infinitesimal mechanism left, which is the relative rotation of adjacent simplexes (any pair). It can be removed by introducing an attaching cable in the tensional layer to link any pair of adjacent simplexes, but introducing more attaching cables can reduce internal forces and increase stiffness significantly (Figure 3.11). In the figure, the bottom layer is assumed to be the tensional layer.

The characteristics of the configurations made of triangular simplexes are similar. The vertex-and-edge connection of geometrically flexible truncated pyramids (Figure 3.5(b)) results in geometrically rigid configuration but strut density is high with low material efficiency. The vertex-to-vertex connection of geometrically flexible tensegrity prisms produces geometrically flexible configuration (Figure 3.12(a)). In order to form geometrically rigid

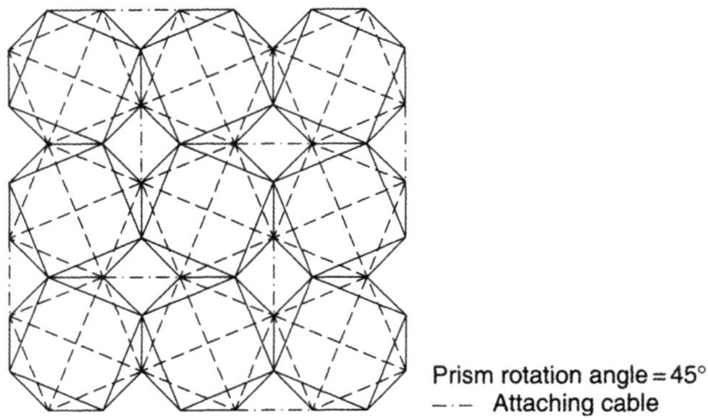

Figure 3.11 Geometrically rigid form for all layouts of contiguous strut configuration Method I for the square prism.

Source: Wang and Li (2001); Courtesy: Multi-Science Publishing.

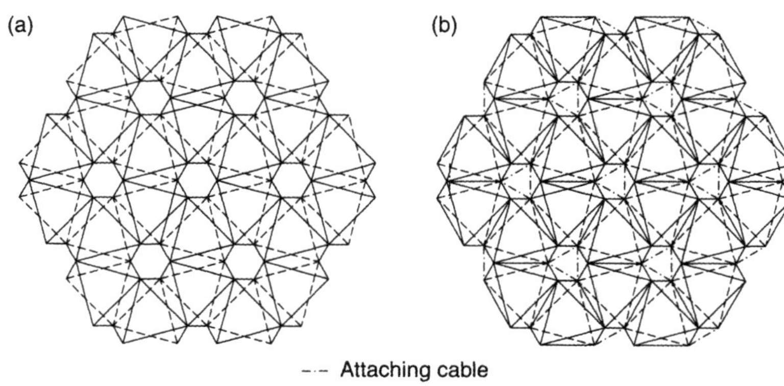

Figure 3.12 Layouts of contiguous strut configuration Method I for the triangular prism: (a) geometrically flexible form; (b) geometrically rigid form (prism rotation angle = 30°).

configuration, stiffened simplexes are required, and an attaching cable is also needed to remove the infinitesimal mechanism of relative rotation of adjacent simplexes, but more attaching cables are introduced in the tensional layer to improve structural properties (Figure 3.12(b)).

3.3 Design examples

Till recently, only a few studies have been carried out on real-scale prototypes, except for some preliminary analysis. That is because few people ever doubted the efficiency of tensegrity concept, since cables are employed as tensional materials and all bars are in compression without redundancy. However, design results turn out to be quite negative. In this section, properties of various tensegrity grids are presented through design examples.

3.3.1 *Introduction*

Hanaor (1994, 1997) presented a real-scale study of a flat tensegrity layout based on the triangular simplexes. The tensegrity grid is also compared with a square-on-square offset space grid. Some important conclusions are drawn. First, the geometrically rigid grid is much stiffer than the geometrically flexible grid with deflection about half whereas the weight saving is apparent although not significant, about one-sixth (note that under the same stiffness requirement, geometrically flexible form would be much heavier). Second, the self-weight of the geometrically rigid tensegrity grid is nearly twice that of the studied space grid (although the space grid is not the optimal form, e.g., it can be improved by supporting at the upper layer with reduced modules). Finally, long bars, a feature of all tensegrity structures, is pointed out as the reason for the heavy weight of tensegrity grids. In fact, there are also other important factors related to the weakness of non-contiguous strut tensegrity grids (Wang and Li 2001), which are adapted in this chapter.

The design of all grids in this book is based on Chinese code (note that the examples are comparative and thus code independent) and internal forces and displacements are computed by the Newton iteration method (Section 2.3). The load factor for dead load is 1.2, for live load 1.4 (so the internal forces are factored). The design strength equals to yield strength divided by material factor 1.1 (but in general, we use smaller value: 200MPa for A3 steel (yield strength = 235MPa), and 300MPa for high-strength 16Mn steel (yield strength = 350MPa). Various tube cross-sections popular in Chinese market are introduced in the design and the selection of these tubes is automatically realized based on full stress design method and the principle that fewer types are preferred. The computation process is based on the following procedure:

(a) Input the initial cross-sections of components.
(b) Compute the form-finding process if prestress is applied.
(c) Compute the load response.
(d) Select the cross-sections of components. If the change of cross-sections is greater than 5%, return to (b).

In general, the iteration of cross-sections needs only 3–5 circles. The resulting cross-sections can satisfy the accuracy requirement in practical constructions.

3.3.2 *Non-contiguous strut tensegrity grids*

In this section, geometrically rigid non-contiguous strut grids formed by the square tensegrity prisms are studied in two-way spanning (in one-way spanning, tensegrity grids present even worse properties; see Appendix A for reference). Although both connection methods are studied, only the results of the edge-to-edge connection (Method II) are listed because the vertex-to-edge connection (Method Ib) is proved less efficient by test computations. The span is prescribed 30m (square plan), as a representative of the familiar range. These grids are simply supported at the bottom boundary nodes. The grid for Layout A is given in Figure 3.13 for reference. Parameters for design are presented as follows.

Span: 30m
Number of modules: 10 for each way
Grid depth: 3m
Dead load: 50kg/m^2 (excluding self-weight of the grid structure)
Live load: 50kg/m^2
Design strength for struts: 200MPa (A3 steel)
Tubes for designing struts: D114t4 (diameter = 114mm, thickness = 4mm), D140t4.5, D159t5, D168t6, D180t7, D194t8, D194t9
Design strength for cables: 500MPa (tendon)
Cross-sections for designing cables (cm^2): 0.5, 1, 2, 3,...

Prestress is not introduced, unless specified. The design depth is relatively large in that the stiffness of non-contiguous strut configurations is low as a whole. Test computations prove that double-layer tensegrity grids are insensitive to non-symmetric loads, such as dead load + half-span live load, so only the full load case is presented. In each grid, several types of tubes and cables are selected from the available list given here. The selection follows the principle that fewer types are preferred (e.g. in the present design, tube types vary from two, when bar forces are evenly distributed, to six, when bar forces are not evenly distributed). The chief results are presented in Table 3.1.

In Layouts A and B, the prism rotation angle is variable. For both layouts, the proportion of slack cables decreases with the increase of the prism rotation angle. In Layout A, the 67.5° case is of much lower internal forces and higher stiffness than the 45° case. But in Layout B, the 67.5° case is less advantageous than the 45° case and its struts are longer. Both the 45° case in Layout A and the 67.5° case in Layout B suffer from uneven distribution of internal forces.

The comparison of three layouts shows that Layout A is the lightest. By inspection, the only difference among three layouts is the arrangement of struts. In Layout A, struts are orthogonal to the edge, whereas other configurations contain struts diagonal to the edge (*it seems that torsional effect*

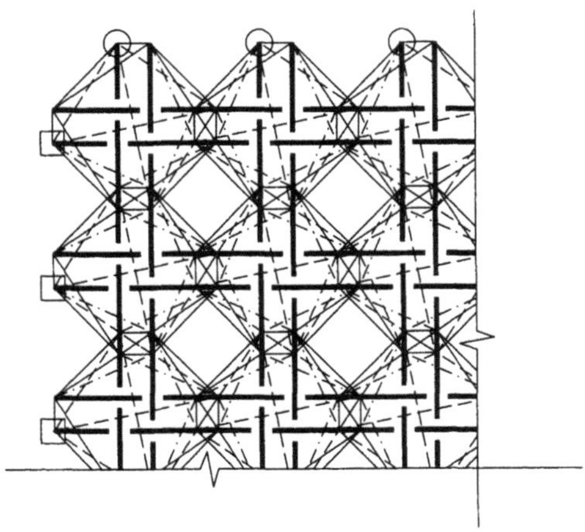

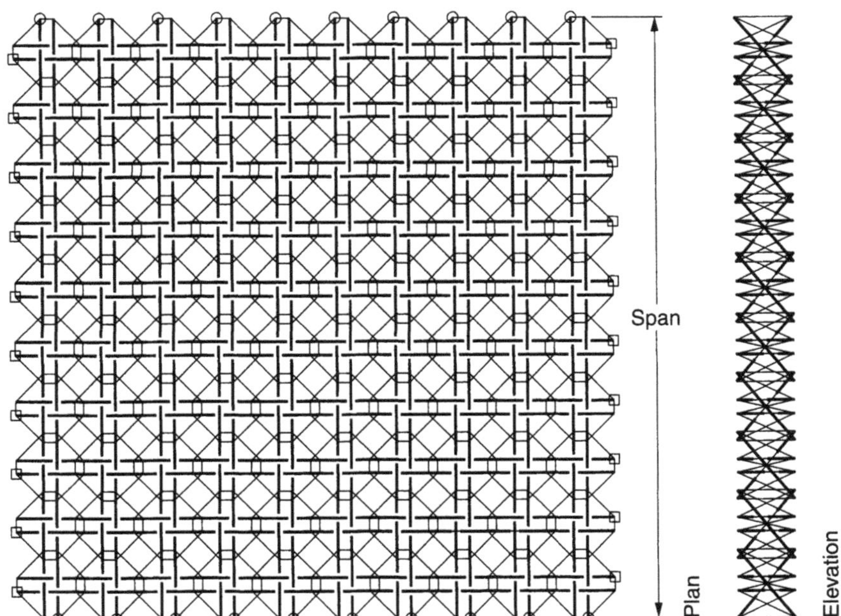

Figure 3.13 Optimal non-contiguous strut grid – Method II, Layout A, 67.5°.
Source: Wang and Li (2001); Courtesy: Multi-Science Publishing.

does not take effect in non-contiguous strut grids). Perhaps owing to the same reason, the layout is also lighter than the vertex-to-edge connection.

In order to study the influence of prestress, same prestress value is applied to all struts for the optimal case – the 67.5° case of Layout A (marked as

Table 3.1 Design results for non-contiguous strut grids of the edge-to-edge connection (Method I)

Layout	*Prism rotation angle*	*Maximum forces under designed load (10kN)*		*Self-weight (Kg/m²)*		*Proportion of slack cables*	*Maximum deflection under live load (span)*	*Strut length (m)*
		Compression	*Tension*	*Struts*	*Cables*			
A	45°	76	53	30.5	8.2	860/2880	1/46	4.646
	67.5°	30	27	28.8	8.2	748/2880	1/71	4.529
	67.5°+ *p*							
	5[a]	30	27	30.3	8.9	647/2880	1/84	4.529
	10[a]	32	28	32.2	9.9	555/2880	1/109	4.529
B	45°	29	23	33.8	9.9	848/2880	1/64	4.646
	67.5°	55	33	36.9	9.5	740/2880	1/57	5.016
C	45°	75	58	33.5	9.5	840/2880	1/41	4.646

Source: Adapted from Wang and Li (2001).

Note
a Prestress value for all struts (unit: 10kN).

'67.5° + P'). The level is 1/6 (5kN) and 1/3 (10kN) of the maximum compression (30kN) under design load respectively. Results in Table 3.1 show that the number of slack cables decreases with the increase of prestress level and the stiffness is improved accordingly. Results also imply that prestress does not increase self-weight significantly and does not have much influence on internal forces when prestress level is not high in geometrically rigid grids.

3.3.3 *Contiguous strut tensegrity grids*

In this section, contiguous strut tensegrity grids formed by the square simplexes are studied under two-way spanning (one-way spanning cases are presented in Appendix A for reference). Two sample grids, Method I, Layout A (67.5°) and Method II, Layout A, are illustrated in Figures 3.14, 3.15, respectively. The design follows the procedure in Section 3.3.2. Parameters different from those in Section 3.3.2 are presented as follows:

Grid depth: 2.5m
Available tubes for designing bars: D76t3.8, D89t4, D114t4, D140t4.5, D159t5, D168t6, D180t6, D194t6, D219t6

Design results show that in Method II, Layout B is less efficient compared with Layout A (Table 3.2). That is because at each inner joint in the upper layer of Layout B (Figure 3.6b), the pair of connected struts includes an inclined angle in plan, resulting in extra force component for the co-planar inclined cable to equilibrate, and the resulting tension in return increases strut compressions and thus strut weight. It follows that the pair of connected

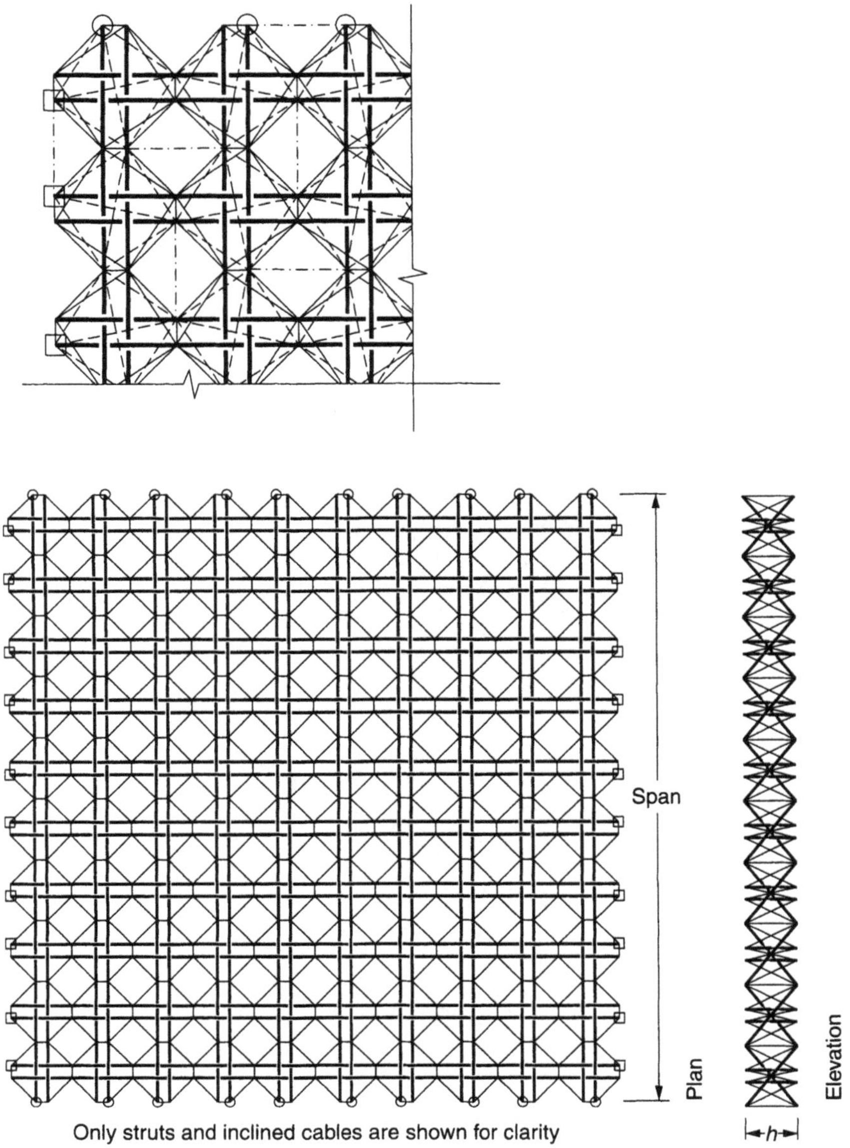

Figure 3.14 Contiguous strut tensegrity grid – Method I, Layout A, 67.5°.
Source: Wang and Li (2001); Courtesy: Multi-Science Publishing.

struts had better be designed 'co-linear' in plan, like the case in Layout A. Owing to the same reason, in Method I, Layouts B and C are much less efficient than Layout A as the included angle is much smaller, so only the results of Layout A (45° and 67.5° cases) are presented.

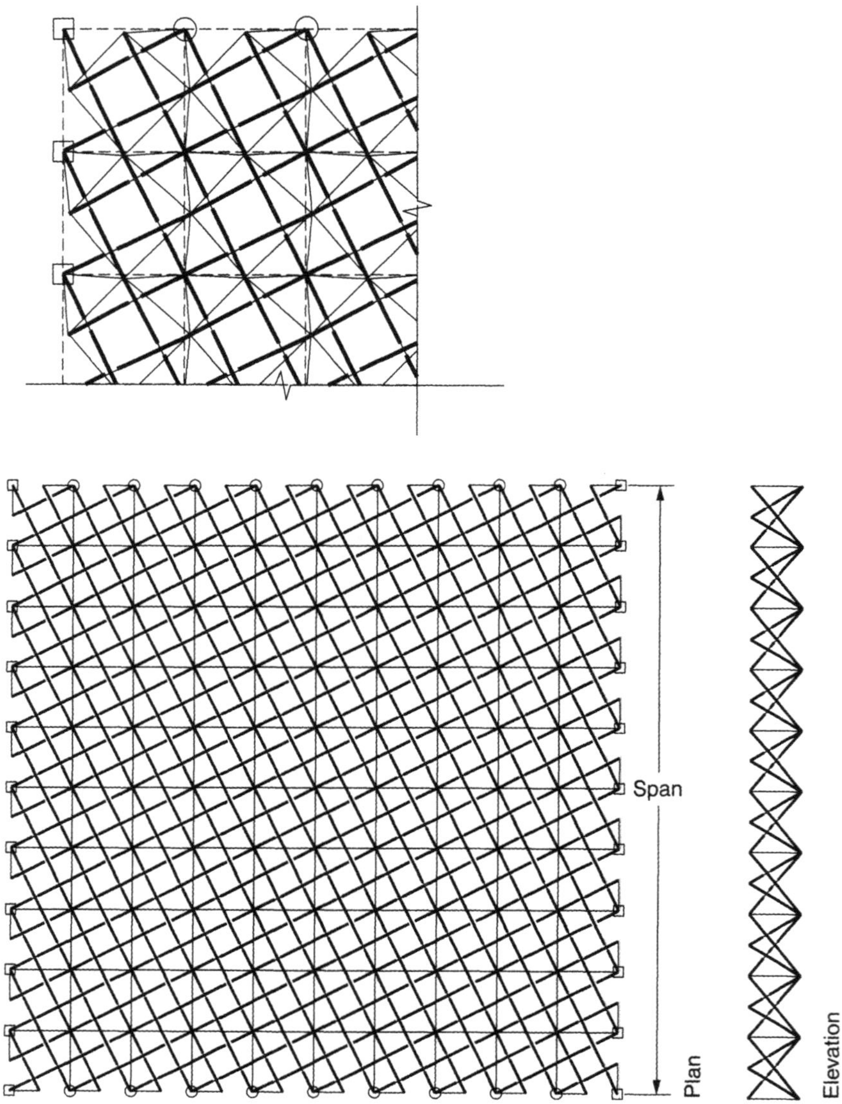

Figure 3.15 Optimal contiguous strut tensegrity grid – Method II, Layout A.
Source: Wang and Li (2001); Courtesy: Multi-Science Publishing.

In Method I, Layout A, deflections under live load, a principal factor for service design, are about 1/180 span. The case of larger prism rotation angle (67.5°) is slightly better. Maximum cable tensions in Method I are roughly half those of Method II at the expense of introducing crossing cables in the tensional layer.

Table 3.2 Design results for contiguous strut grids

Methods and layout	*Prism rotation angle*	*Maximum forces under designed load (10kN)*		*Self-weight (Kg/m²)*		*Proportion of slack cables*	*Maximum deflection under live load (span)*	*Strut length (m)*
		Compression	*Tension*	*Struts*	*Cables*			
I A	45°	19	13	16.9	4.0	422/1719	1/179	3.905
	67.5°	18	15	16.2	3.7	410/1719	1/184	
II A	45°	18	29	16.9	2.9	360/1020	1/184	4.183
B		22	29	19.6	3.8	340/1020	1/134	

Source: Adapted from Wang and Li (2001).

By comparing two types of connection methods based on the optimal cases (Layout A, 67.5° case for Method I and Layout A for Method II), the former has more cables and connections, and the latter is subject to slightly longer struts. They are roughly identical in self-weight and stiffness, but in practical construction the latter is better.

3.3.4 *Comparison of optimal tensegrity configurations*

In this section, the optimal cases of contiguous strut tensegrity grids are compared with the optimal non-contiguous strut tensegrity grid, and a square-on-square offset space grid (SOS grid), one of the popular space truss forms, in 30m span. The optimal cases for tensegrity grids are Method II, Layout A, 67.5° for non-contiguous strut grids (Figure 3.13); Method I, Layout A, 67.5° (Figure 3.14) and Method II, Layout A (Figure 3.15) for contiguous strut grids. The conditions are the same as those in Section 3.3.3 (note that grid depth is 2.5m for all).

Note that the optimal non-contiguous strut tensegrity grid and the contiguous strut grid of Method I, Layout A follow the same orientation of struts and that the only difference between them is just the configuration method, thus the efficiency of two types of configurations can be compared straightforwardly. The contiguous strut grid of Method II, Layout A with simpler geometry can be compared with the space truss. The results are presented in Table 3.3. Because the studied conditions are typical in design, the results are considered representative of the properties of various grid types.

In order to see their properties more clearly, internal forces distribution of tensegrity grids (excluding web cables) and the space truss (excluding web bars) are illustrated in Figures 3.17–3.19, respectively. Most upper cables in the non-contiguous strut tensegrity grid do not slacken and large part of them is subjected to relatively large tensions. Whereas in contiguous strut tensegrity grids, most upper cables slacken, and a few subjected to small tensions mainly lie in boundary modules. Note that the distributions are common for all tensegrity grids.

Table 3.3 Comparison of optimal tensegrity grids and space truss

Grid types	*Max. forces under design load (10kN)*		*Self-weight (Kg/m²)*		*Max. deflection under live load (span)*	*Strut length (m)*	*Tube cross-section (D × t)*	*Cable cross-section (cm²)*
	Compression	*Tension*	*Bars*	*Cables*				
Optimal non-contiguous strut grid	34	33	26.4	8.5	1/66	4.215	89 × 4 114 × 4 140 × 4.5 159 × 5	0.5, 1, 2, 3, 4, 5, 7
Contiguous strut grid – Method I, Layout A, 67.5°	18	15	16.2	3.7	1/184	3.905	76 × 3.8 89 × 4 114 × 4 140 × 4.5	0.5, 1, 2, 3
Contiguous strut grid – Method II, Layout A	18	29	16.7	3.1	1/184	4.183	76 × 3.8 89 × 4 114 × 4 140 × 4.5	0.5, 1, 2, 3, 4, 5, 6
Square-on-square offset space grid	14	13	14.0	—	—	3 3.279	48 × 3.5 60 × 3.5 76 × 3.8 89 × 4 114 × 4	—

Source: Adapted from Wang and Li (2001).

Design results show that non-contiguous strut grid is much larger in internal forces, weight and deflection than contiguous strut grids, so are contiguous strut grids than the space truss except for the deflection aspect due to different material application. The inherent reason is explained in the following section.

3.4 Structural efficiency of tensegrity grids

The invention of novel structural systems of cables and struts with higher structural efficiency can be feasible based on clear understanding of the low structural efficiency of tensegrity grids.

- Structural efficiency is defined by the reverse of the weight of the grid specified to be capable of sustaining the prescribed loading conditions and satisfying service requirements.

Clearly, the higher the weight, the lower is the structural efficiency. The chief service requirement is stiffness (deflection control). A structure of low stiffness requires high prestress to meet service requirements, thus internal forces and consequently, self-weight is increased. Properties related to the structural efficiency of two tensegrity configurations are presented as follows and then summarized.

3.4.1 Properties of contiguous strut tensegrity grids

The properties of non-contiguous strut tensegrity grids are more complex than those of contiguous strut grids, so the latter are explained in advance.

Internal forces in the structural components in either tensegrity grids or space trusses (both belong to space frames as a whole) arise from the action of bending moments and shear forces on the structure as a whole. Of these two actions the effect of internal moments is dominant. The internal forces of components forming the compressive/tensional layer are determined by the corresponding resistant level arms to the tensional/compressive centre of the cross-section (in the context, the resistant level arms). The deeper the level arm, the smaller is the force (for further discussion refer to Section 5.1.1).

In contiguous strut tensegrity grids, upper cables slacken or sustain small tensions following the 'compressive' deformation of the upper layer. Therefore, internal moments are sustained by inclined struts and bottom cables whereas the influence of upper cables is negligible as a whole. The internal forces of contiguous strut grids are generally larger than those in space trusses owing to the reduced resistant lever arms as a result of bar inclination. The resistant lever arms of inclined struts in resisting internal moments are about 80% of the grid depth for the studied parameters, and the lever arms of bottom cables can be estimated referring to the case that struts are joined at centres, that is, about half the grid depth (Figure 3.16(a)). In contrast, both lever arms in space trusses are the full depth of the grid. It explains larger internal forces in tensegrity grids, which can refer to their difference in force distributions in Figures 3.17 and 3.18. Larger strut compressions and especially larger cable tensions in connection Method II (Figure 3.17(a)) are obvious as compared with the internal forces in the space truss (Figure 3.17(b)). The case is similar for Method I except that large cable tensions are not explicit (up to 77kN, Figure 3.18) due to much higher cable density (7:2 compared with the bottom chords in the space truss).

In addition to the reduction in the resistant lever arms, another factor, which is more important, is obviously their excessive bar length hence larger tube cross-sections are designed. Consequently, the self-weight of the optimal tensegrity grids is about 40% heavier than that of the space truss even under higher grade for cables (Table 3.3). The difference in fact becomes larger with the increase of span.

By the way, contiguous strut configurations with openings (or called 'plane-filling forms') are of low structural efficiency owing to the resulting isolation of struts, which results in cables sustaining tension in the compressive layer. The properties are partially similar to those of non-contiguous strut grids, as can be analysed as follows.

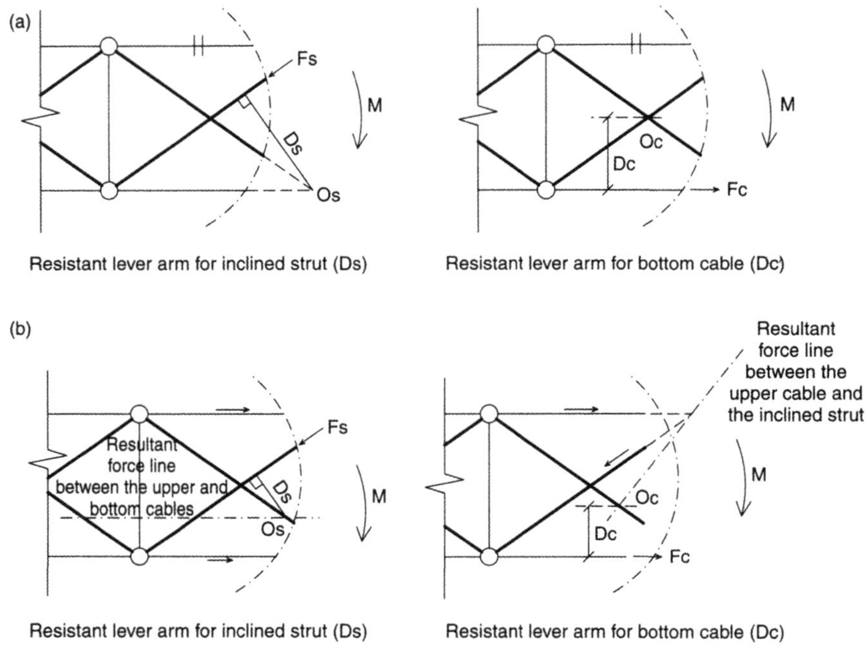

Figure 3.16 Resistant lever arms in tensegrity grids: (a) contiguous strut grid; (b) non-continuous strut grid.

Source: Adapted from Wang and Li (2001).

3.4.2 Properties of non-contiguous strut tensegrity grids

From analytical results, non-contiguous strut tensegrity grids are characteristic of large internal forces, very low stiffness and heavy weight and are actually sensitive to support positions. The deformation and internal forces of upper cables, inclined struts, and bottom cables of the studied optimal case are illustrated in Figure 3.19. Among factors leading to their low efficiency, excessive bar length and the reduction of the resistant lever arms due to bar inclination are the same as in contiguous strut grids. But *the dominant factor is actually the strut-to-cable connection among simplexes* as can be inferred from the comparison between the contiguous type of Method I, Layout A and the non-contiguous type (Table 3.3). It is the fundamental reason why the properties of non-contiguous strut tensegrity grids are much poorer in structural efficiency than contiguous strut grids.

The strut-to-cable connection causes low efficiency in that the internal forces among simplexes are transferred indirectly through joints to cables. Consequently, cables in the compressive layer (here, upper cables) always remain in tension and infinitesimal mechanisms occur at the connected vertices. Additional attaching cables can 'remove' these mechanisms in view of

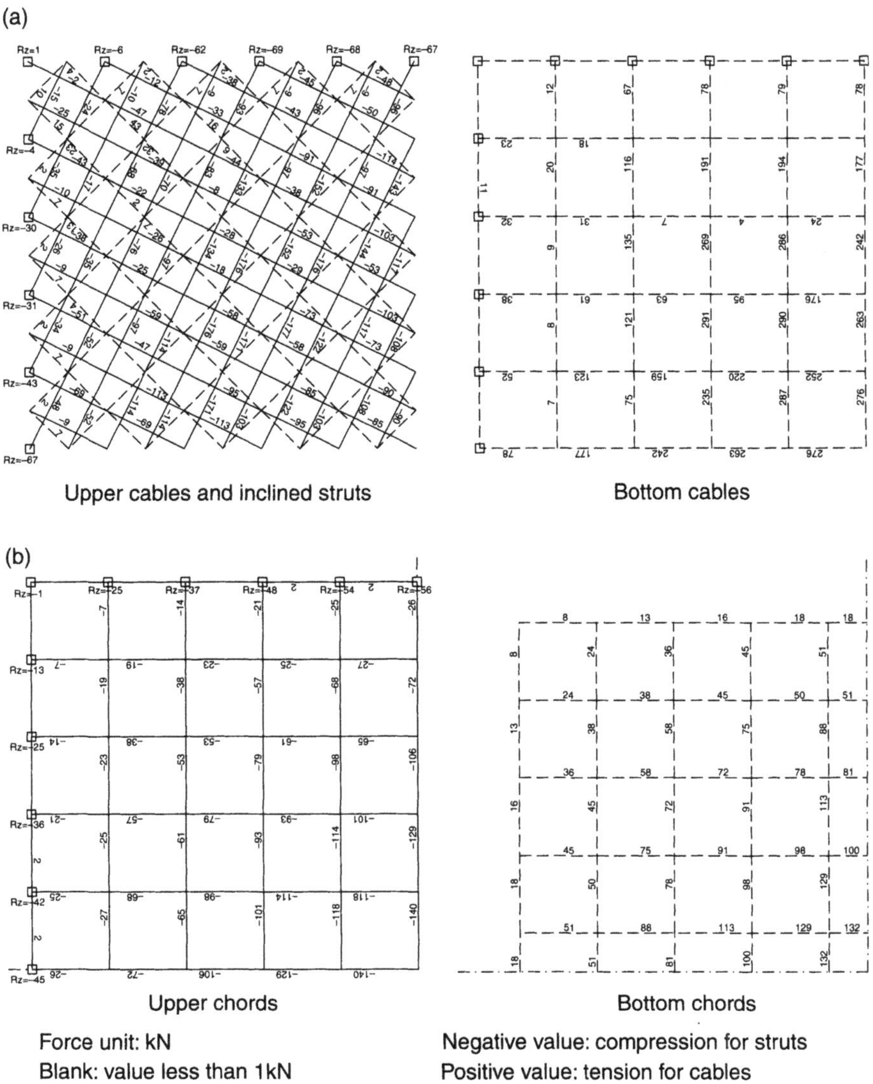

Figure 3.17 Internal forces for the optimal contiguous strut tensegrity grid and SOS grid: (a) optimal contiguous strut tensegrity grid (quarter layout, rotational symmetry); (b) SOS grid (quarter layout).

Source: Wang and Li (2001); Courtesy: Multi-Science Publishing.

geometry, forming geometrically rigid form, but considerable part of them actually still exists. As illustrated in Figure 3.20, the resultant forces at the connected vertices from adjacent simplex are compression in the upper layer and tension in the bottom layer. In the bottom layer the tensions can

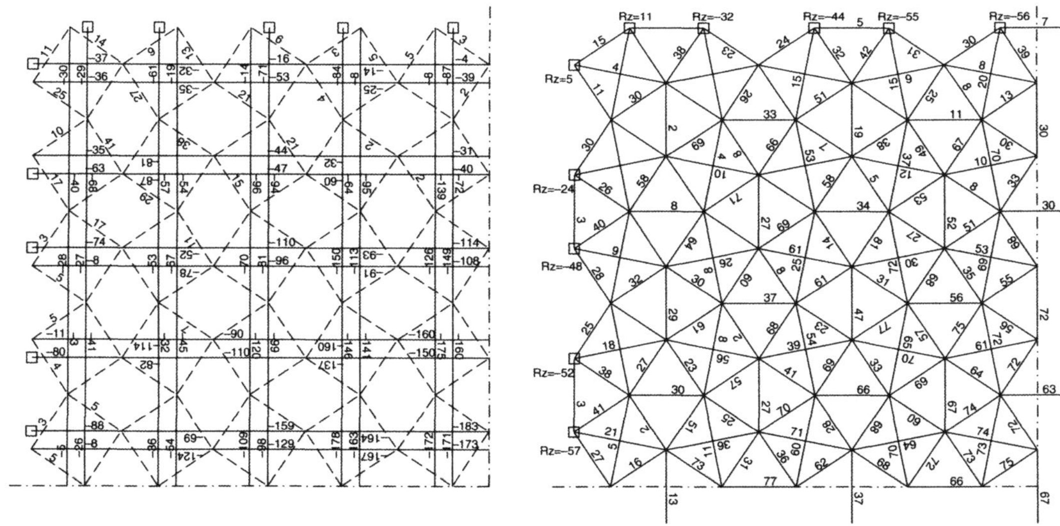

Figure 3.18 Internal forces for the contiguous strut tensegrity grid – Method I, Layout A, 67.5° case.

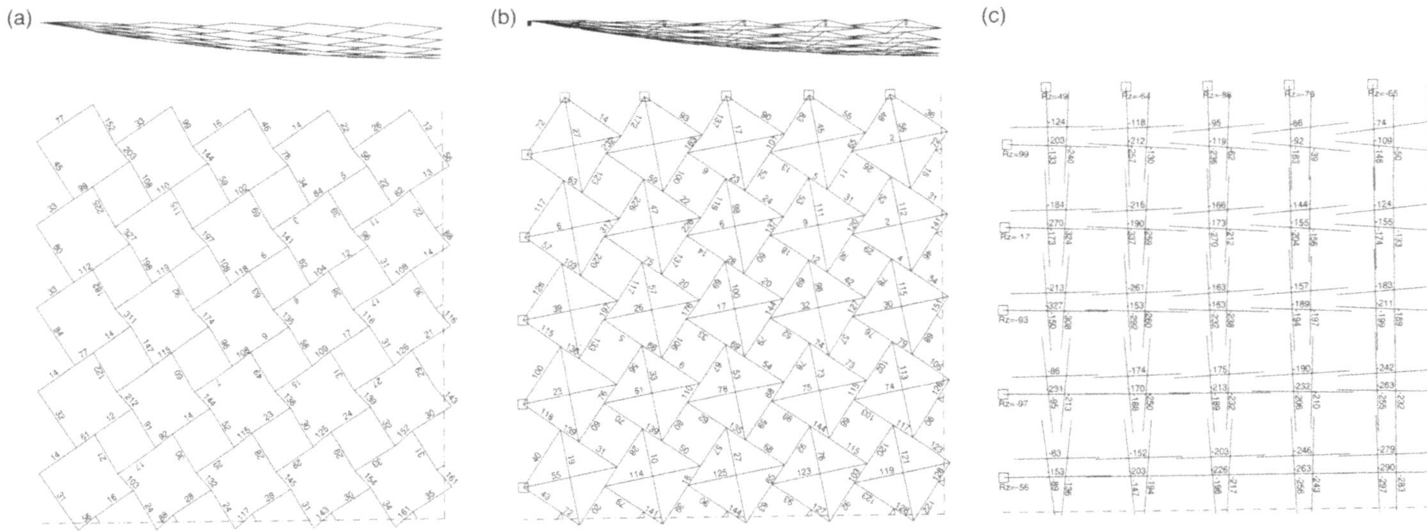

The deformation and forces are of rotational symmetry – one quarter of the grid (which is shown) rotating every 90° presents the results of the left quarters.

Force unit: kN
Blank: value less than 1 kN

Negative value: compression for struts
Positive value: tension for cables

Figure 3.19 Deformation and internal forces for the optimal non-contiguous strut tensegrity grid (quarter layout): (a) upper cables; (b) bottom cables; (c) inclined struts.

Source: Wang and Li (2001); Courtesy: Multi-Science Publishing.

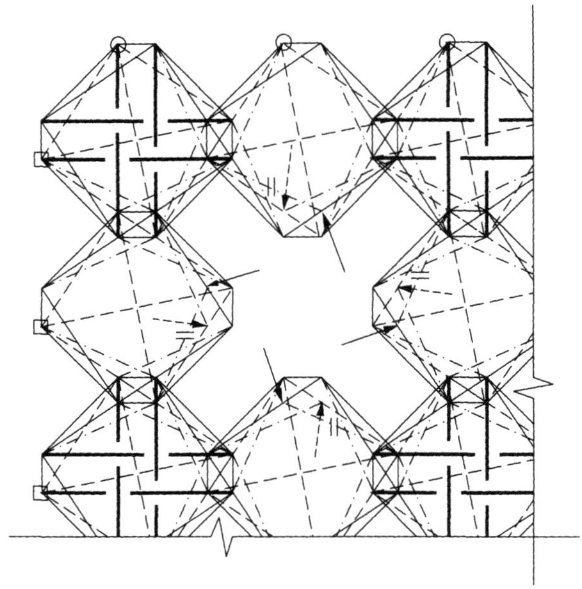

—▸ Compressive resultant transferred from adjacent simplex

- - -▸ Tensional resultant transferred from adjacent simplex

|| Attaching cable inclined to slacken under load

Figure 3.20 Analysis of mechanism geometry under load (central simplex taken out, for detailed drawing see Figure 3.13).

generally be balanced by the corresponding attaching cables, so most mechanisms can be avoided and the deformation is small as a whole (Figure 3.19(b)). However, in the upper layer, large part of infinitesimal mechanisms will be activated as the attaching cables are inclined to slacken or share very small forces (these infinitesimal mechanisms are thus 'implicit', as mentioned in Section 2.4.2). It explains the large deformation and internal forces in the layer, as verified in Figure 3.19(a). (The cases with camber, which may refer to Appendix B, are better in that forces are much reduced but do not present significant improvement in structural efficiency because near-mechanism geometry is still unavoidable.)

Upper cables in high tension results in additional significant reduction in the resistant lever arms besides bar inclination by offsetting the moment resistance of the bottom cables and inclined struts, respectively. The concept of the reduction in resistant lever arms for non-contiguous strut grids is illustrated in Figure 3.16(b), compared with that for contiguous strut grids (Figure 3.16(a)). The increases in strut compressions in return enlarge tensions in upper cables. The developed large deformation and tension are then transferred to all other components.

3.4.3 Summary of structural efficiency

Tensegrity grids are not structurally efficient despite that high-strength cables are introduced as tensional material and that all bars are in compression as they do not comply with the dominant load-transfer pattern. Due to the low structural efficiency, tensegrity grids are suitable in small spans or when special architectural requirements become the principal concern (as the example, large span applications of non-contiguous strut grids are discussed in Appendix B) or in special functions such as deployment, as discussed in Chapter 6. Based on the present analysis (The present discussion is focused on flat forms, excluding domical forms relying on lateral supports. But conceptually, the property should be similar), the low structural efficiency can be summarized into three factors, which are ranked in descending order according to the significance:

- Isolation of struts in grid,
- Excessive bar length, and
- Reduced resistant lever arm owing to bar inclination.

The latter two factors are due to the *isolation of struts in simplex*, which is common for all tensegrity grids.

Isolation of struts in grid

In non-contiguous strut tensegrity grids, struts are isolated among simplexes. The indirect force transfer leads to cables in tension in the compressive layer and infinitesimal mechanisms (or near-mechanism geometry) that enlarge the tensions, resulting in much-reduced resistant lever arm and low stiffness. The drawback is dominant and is conceptually common for all non-contiguous strut grids, including those that are not composed of simplexes. In addition, 'isolation of struts in grid' increases significantly the number of joints that require much more components to restrain the degree of freedom and results in complexity in geometry especially when curved forms are required.

As improved forms, contiguous strut tensegrity grids avoid the 'isolation of struts' in simplex connections and thus present much better structural efficiency over non-contiguous strut tensegrity grids. Meanwhile, as struts are connected at ends, large part of contiguous strut configurations can directly realize geometrical rigidity. It follows that structurally efficient grids should be at least *based on contiguous strut configurations*.

Isolation of struts in simplex

In tensegrity grids struts are isolated in simplexes. These struts are required to stabilize a tensional outer space without contacting each other; thus struts are attached to far vertices of simplexes. It results in long bars subject to buckling and bar inclination that causes the reduction of the resistant lever arms for struts and especially cables. Hence, the structural efficiency of contiguous strut tensegrity grids is still significantly lower than that of space trusses. So if we expect that the resulting grids can be structurally efficient, *struts should be allowed to be in contact in simplexes.*

3.4.4 How to achieve high structural efficiency

From structural point of view, tensegrity concept is the introduction of 'isolation of compression members (struts)' in free-standing cable networks. However, based on the analysis above, low structural efficiency of tensegrity concept stems exactly from the introduction of 'isolation of struts' in simplexes and simplex connections. It leads to the conclusion that only contiguous strut configurations made of simplexes containing contiguous struts could present high structural efficiency.

The defect of tensegrity concept in structure leads to the search for lightweight free-standing grids of cables and struts by discarding the concept of 'isolation of struts' in both simplexes and simplex connections. Consequently, the discussion of structural efficiency leads to the invention of lightweight forms that are more 'reasonable' from structural point of view, and in addition, present new architectural art.

Part II

Cable-strut systems

4 Geometrical characterization of basic cable-strut systems

Tensegrity structures are low in structural efficiency because of the introduction of the concept 'islands of compression in a sea of tension' in free-standing grids. Although the concept is undoubtedly an invention in architecture, it is not sufficient for successfully creating lightweight structures as the original objective. Lightweight free-standing grids made of cables and struts can be invented successfully through avoiding the drawbacks and inefficient forms of tensegrity grids. The nature of tensegrity grids as dominantly composed of simplexes leads naturally to the idea that improved structural forms should also be based on the innovation in simplexes, either by new methods to stabilize the original tensegrity building blocks or by designing novel simplexes. Consequently, tensegrity grids are extended to cable-strut grids. The concept was discussed preliminarily by the author (Wang 1998a,b) and studied systematically afterwards (Wang and Li 2001, 2003a,b, Wang 2002a,b). Geometrical details are presented in this chapter.

4.1 Definition of cable-strut systems

Cable-strut systems are extended from tensegrity systems. The term 'cable-strut' itself contains the following meaning:

- Cables as tensional material. 'Cable' here is the generalized term for cables or tendons, such as steel strands, steel wires and tension rods, etc. that are either in tension or inactive. It is utilized to reduce weight by the high-strength lightweight property, and in addition, to simplify joint connection.
- Bars only as compressive material. Bars in tension are only occasional (thus cables are normally 'continuous'). The specification makes sure that bars can be saved to the ultimate degree so as to reduce weight. That is the reason why the systems are termed '*cable-strut systems*', not the more generalized '*cable-bar systems*'. In contrast, bars in conventional bar systems are subjected to either compression or tension or in general, compression and tension alternatively under various load cases.

- The whole structure is stabilized by the interaction of cables and struts only. That is to say, there are no boundary anchoring systems. Accordingly, cable domes and radial cable roofs are *cable-strut – ring beam* systems.

Based on the spirit above, cable-strut systems can thus be defined:

- Cable-strut systems are free-standing pin-jointed systems of struts and continuous cables.

Here, 'continuous cables' means cables are inter-connected to be continuous in geometry, otherwise, bars in tension is conceptually unavoidable. In comparison with the definition of tensegrity systems, cable-strut systems remove the restriction of 'isolation of struts'. The extended systems cover a wide variety of forms, including ornamental forms. However, studies in this book are focused primarily on structural forms – cable-strut grids. Cable-strut grids are in general composed of cable-strut simplexes and the configurations include non-contiguous strut types and contiguous strut types.

4.2 Principles of designing cable-strut simplexes

The basic principle of designing cable-strut simplexes, evidently, lies in how to realize weight reduction. The weight of the resulting grid is contributed by struts, cables and joints. Improved resistant lever arms and less slack cables are preferred for the reduction of cable weight. Improved resistant lever arms can also reduce joint weight, and reducing joint density and avoiding small included angles among components are related design techniques. However, the main concern here is the minimization of strut weight, which is the main index introduced in simplex design.

4.2.1 *Evaluation of strut weight*

Whether a cable-strut grid is lightweight can be evaluated when compared with other grids. Under the same layouts and loading conditions, the strut weight of various grids can be compared approximately by the following function:

$$\sum V_i = \sum A_i L_i, \qquad i = 1, n \tag{4.1}$$

in which, n is the number of struts per area (strut density), V_i, A_i, L_i being the volume, cross-sectional area and length of a strut i, respectively.

The cross-sectional area A_i of a strut subjected to buckling can be expressed as follows:

$$A_i = C \cdot F_i \cdot S_i^2 \tag{4.2}$$

in which, C is a constant of material, F_i and S_i is the internal force and slenderness of a strut i, respectively. In actual structural design, A_i is subject to the restriction of smallest cross-section and S_i to the smallest slenderness (in China, 180).

The substitution of Eqn (4.2) into Eqn (4.1) yields

$$\sum V_i = C \cdot \sum F_i \cdot S_i^2 \cdot L_i, \qquad i = 1, n \tag{4.3}$$

Eqn (4.3) expresses generally the influence of various factors on the total volume or equivalently, weight of all struts. Therefore, general principles of weight reduction can be obtained.

4.2.2 *General principles*

In Eqn (4.3), the item of slenderness squared expresses the influence of the length of struts (L_i) in buckling effect. The length of struts (L_i) is undoubtedly the primary factor affecting the weight of struts and hence the whole structure. Therefore, small bar length becomes a primary concern in inventing novel cable-strut simplexes. As in cable-strut simplexes struts are allowed to be in contact, it gives the freedom in designing short struts.

Bar forces (F_i) in contiguous strut grids are determined chiefly by their resistant lever arms when modular parameters are given. Cable-strut simplexes should allow for relatively high resistant lever arms in grid including optimal design. Then the factor becomes less influential among structurally efficient forms (Chapter 5).

The strut density (n) is determined by the number of struts in each simplex and the way of connecting simplexes for the given modular length. As fewer struts are preferred, cable-strut simplexes are generally of simple geometry.

Polyhedra of simple geometry include prisms, anti-prisms, pyramids, truncated pyramids, anti-truncated pyramids and reciprocal prisms (di-pyramids), etc. as shown in Figure 4.1(a)–(f). As the building blocks of lightweight grids, cable-strut simplexes can be formed by the aforesaid principles to stabilize these polyhedra by struts and cables. The topic is discussed in detail in the following section.

Irregular polyhedra of simple geometry, such as those with different top base and bottom base in Figure 4.1(g), require further investigation. Higher polyhedra that are not suitable as building blocks for lightweight structures but are capable of forming ornamental forms are excluded from the present discussion.

4.3 Cable-strut simplexes

The narrow sense of cable-strut simplexes includes only cable-strut simplexes that can be applied to form lightweight cable-strut grids whereas the broad sense also includes tensegrity simplexes. In cable-strut simplexes,

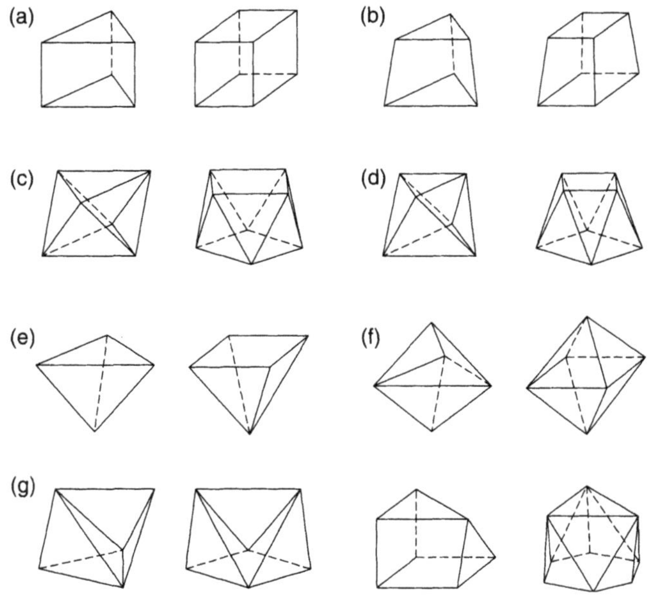

Figure 4.1 Polyhedra of simple geometry: (a) prisms; (b) truncated pyramids; (c) anti-prisms; (d) anti-trucated pyramids; (e) pyramids; (f) reciprocal prisms (di-pyramids); (g) irregular polyhedra of simple geometry.

struts are allowed in contact, giving the freedom in designing short struts and/or reducing strut number. In order that all bars are in compression and that bars are of reduced length and density, cable-strut simplexes are in general of simple geometry. Moreover, simple geometry has the benefit that the distribution of internal forces is not so complex in the resulting grid. Cable-strut simplexes can be formed by new methods to stabilize polyhedra that form tensegrity simplexes or by designing new polyhedra. Studies are focused on the triangle-based and square-based simplexes. Square simplexes can be used in building square and rectangular layouts, and triangular simplexes in elliptical and irregular layouts.

4.3.1 *Anti-prisms and anti-truncated pyramids*

Tensegrity prisms have the feature of non-contiguous struts. Each realizes a stable volume of continuous cables and discontinuous struts, that is, 'tensegrity'. When the struts are joined at centres, the volume becomes unstable and thus lateral inclined cables are required to stabilize the simplex (Figure 4.2). The resulting bar-intersecting simplexes with half-reduced bar length belong to a new type of cable-strut simplexes, namely, anti-prisms (APs) and anti-truncated pyramids (ATPs). In contrast, the prism rotation angles in new simplexes can vary randomly between larger than zero and half central

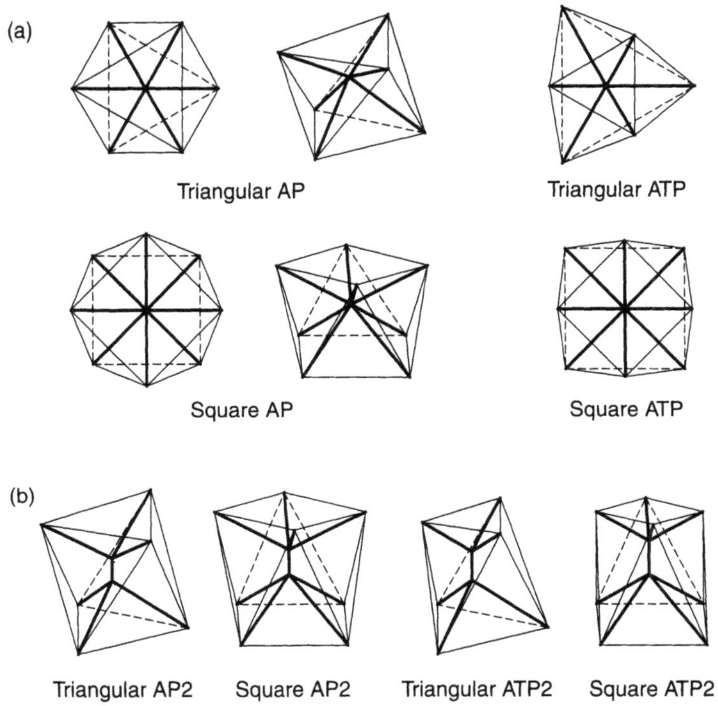

Figure 4.2 AP and ATP simplexes: (a) APs and ATPs; (b) AP2s and ATP2s.

angle (e.g. 60° for the triangular, 45° for the square). In Figure 4.2(a), the angle of each simplex is half the vertex angle. Each AP or ATP is geometrically rigid (note that for the squares, reinforcing the base is not necessary) with three degrees of static indeterminacy based on the initial geometry for both the triangle and the square.

An AP or ATP can also be stabilized by designing two inner joints, as shown in Figure 4.2(b). Such simplexes can be called an AP2 or ATP2. An AP2 (ATP2) appends a vertical strut that connects upper and lower inclined struts. Each AP2 (ATP2) has a state of self-stress and no mechanisms. The introduction of the vertical strut facilitates the deployment as it can be realized easily by telescoping only one strut (the vertical strut) in each simplex, which is discussed in Section 6.3. More inner joints may be designed but the topic is beyond the scope of this book.

The so-called AP or ATP simplexes include all designs but in the narrow sense may refer to the ones with one inner joint.

4.3.2 *Prisms and truncated pyramids*

A prism (P) is topologically identical to an anti-prism or anti-truncated pyramid. The resulting simplexes can also be formed by short struts from inside and relatively longer edge cables from outside. The principles of forming cable-strut simplexes by truncated pyramids (TPs) are the same. The resulting TP forms can be considered as the variation of P simplexes.

The simplex is called a P when it contains only one inner joint (Figure 4.3(a)). A triangle-based P contains an infinitesimal mechanism of prism rotation and a self-stress state. The only self-stress state can be easily figured out from geometry, as at each joint three cables enclose one strut. Given force to any member determines the whole force state. The only mechanism is obvious from geometry, which can also be got from the

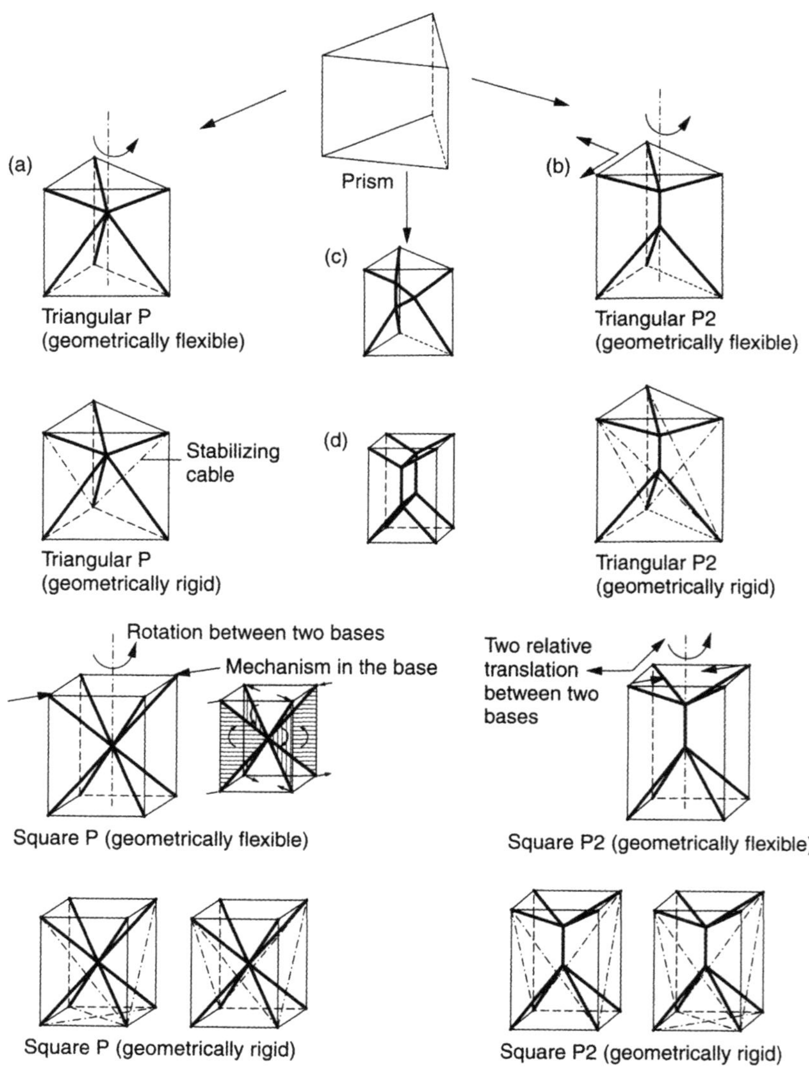

Figure 4.3 P simplexes: (a) Ps; (b) P2s; (c) P3; (d) P4.

Source: Adapted from Wang and Li (2003a).

generalized Maxwell's rule (Eqn 2.1b). In order to achieve better stiffness, the mechanism can be avoided by introducing a diagonal stabilizing cable in a lateral face in view of geometry. However, as it may slacken under rotation, another cable of reverse clockwise shall be added.

A square-based P contains a state of self-stress, a finite mechanism in base and an infinitesimal mechanism of relative rotation between two bases (actually arbitrary two opposite faces). Typically, the prism rotation mechanism can be avoided by a pair of diagonal cables in lateral faces and the finite mechanism can be avoided by introducing a pair of crossing cables in the upper or bottom base (Figure 4.3(a), bottom left). Another choice, as shown in Figure 4.3(a) (bottom right), is that two mechanisms can be avoided by introducing two pairs of diagonal cables.

The simplex is called a P2 when it contains two inner joints (Figure 4.3(b)). Compared with a P, a P2 appends a vertical strut that connects upper-and bottom-inclined struts. The vertical strut introduces two additional infinitesimal mechanisms of translations between two bases. In a triangle-based P2, three infinitesimal mechanisms can be stabilized by four lateral cables. In a square-based P2, it is interesting to find that the introduction of two pairs of lateral cables, just like that in the case of a square-based P, can also stabilize the two additional translations in view of geometry with a state of self-stress (Figure 4.3(b), bottom left). However, in consideration of cable slackening, crossing cables may be required in the base depending on actual forces (Figure 4.3(b), bottom right).

Prisms with more inner joints among possible forms (P3, P4, etc.) are presented in Figure 4.3(c), (d) for reference. Ps P2s and P3s, etc. are called P simplexes generally, but the narrow sense may include only Ps.

4.3.3 *Reciprocal prisms and di-pyramids*

Reciprocal prisms and di-pyramids refer to the same polyhedron, but they are used separately to describe simplexes of different composition. As cable-strut simplexes, a reciprocal prism (RP) is made of a vertical strut (VS), a certain horizontal struts (HSs) and edge cables (ECs) connected by hinged joints (Figure 4.4). The horizontal struts are enclosed, forming the base polygon. In each RP, there is only one state of self-stress, and if no cable slackens there is no inner mechanism. The self-stress state can be easily understood as each joint in the base is connected with only four components. Actually, RPs are

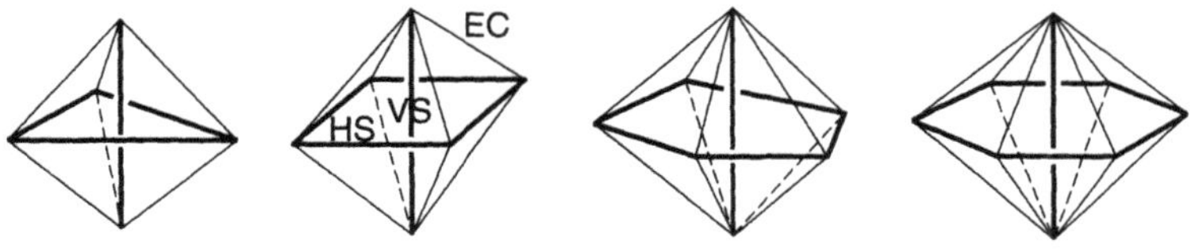

Figure 4.4 RP simplexes.

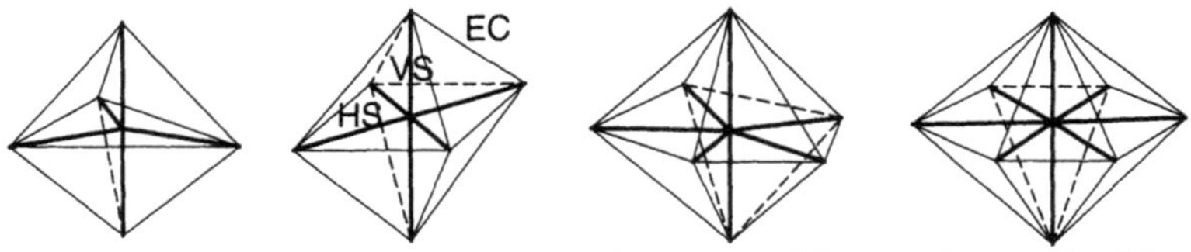

Figure 4.5 DP simplexes.
Source: Wang (2002a).

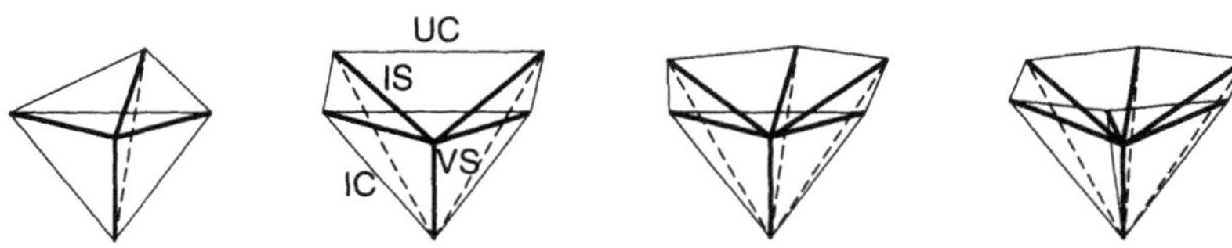

Figure 4.6 CP simplexes.

introduced as the first cable-strut simplexes that are applied to form grids in order to avoid the low structural efficiency of tensegrity grids (Wang 1996a).

As a cable-strut simplex, a di-pyramid (DP) contains struts in the base polygon that are connected at its centre and two perpendicular struts that are connected at the same centre (Figure 4.5). The other end of these struts is connected to edge cables that enclose the simplex volume. From composition, a DP resembles two pyramids combined together, and that is how the name is obtained. The simplexes were patented in China (Wang 2002a). A DP does not contain mechanisms. A triangular DP contains two states of self-stress, and a square contains three, etc.

4.3.4 Crystal-cell pyramids

A crystal-cell pyramid (CP) is made of a vertical strut (VS), a certain inclined struts (ISs) and outer cables connected by hinged joints to stabilize a pyramid (Figure 4.6). Outer cables include upper cables (UCs) and inclined cables (ICs). UCs form the upper base and ICs form the diagonal edges. Each inclined strut is connected to the upper base in one end and to the VS in the other. In each CP, there is also only one state of self-stress and no inner mechanisms if no cable slackens. It was also patented in China (Wang 2002b).

4.3.5 Summary of basic cable-strut simplexes

As pointed out in Chapter 1, the planar form of tensegrity simplexes is a two-strut 'X' module (Figure 1.16). Accordingly, planar (2D) forms of cable-strut simplexes are summarized in Figure 4.7. The planar form of RPs is a two-strut '+' module, that of CPs a three-strut 'Y' module, that of DPs

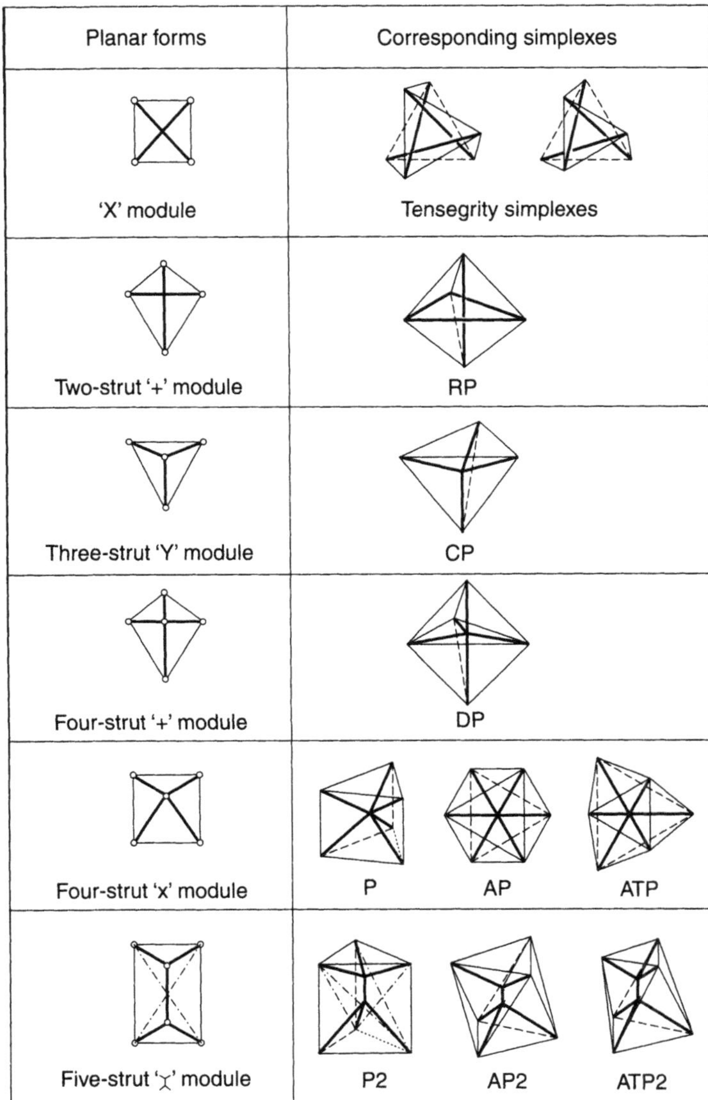

Figure 4.7 Corresponding planar forms of cable-strut simplexes.

a four-strut '+' module, and that of APs, ATPs and Ps is a four-strut 'X' module. Finally, the planar form of AP2s, ATP2s, P2s is of five-strut 'dumb bell' module developed from 'X' module with a pair of stabilizing cables added.

Except for the RP simplexes, all the other simplexes are stabilized by short struts from inside and relatively longer cables from outside, and among them, a CP has the smallest strut density.

4.3.6 Simplex structures

All cable-strut simplexes themselves can be a roof structure. An example in Figure 4.8(a) shows a CP structure with base nodes supported. When struts are long, they may be replaced by super struts (Appendix B). Secondary struts can be added, when necessary, to shape faceted surface, and increased volume can be realized by designing twin simplexes together (Figure 4.8(b)). More studies on simplex structures from architectural point of view are presented in Chapter 7.

It is mentioned that irregular cable-strut polyhedra offer more choices in design. The example in Figure 4.8(c) shows a polyhedron of hexagonal base and an edge top. Membrane covering is to be attached to two large trapezoidal faces.

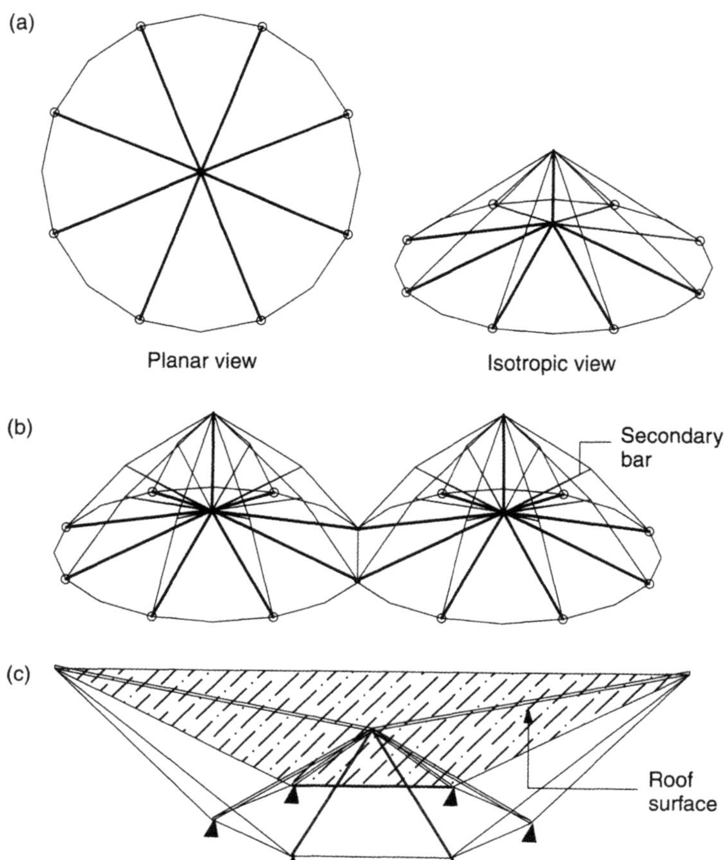

Figure 4.8 Simplex structures: (a) a CP structure; (b) twin-CP structure with secondary bars; (c) a structure of irregular simplex.

4.4 Cable-strut grids of non-contiguous strut configurations

Based on the geometry, cable-strut simplexes can be applied into non-contiguous strut configurations in which struts are not connected directly among simplexes (among them, only RPs seem not suitable as their bases are composed of bars). The resulting grids belong to cable-strut grids in the general sense, and are structurally more efficient than non-contiguous strut tensegrity grids owing to the reduction of strut length. However, these grids are still not structurally efficient due to cables in tension in the compressive layer. Despite that, such a form expresses the value of cable-strut simplexes in architecture, presenting isolation of strut groups while avoiding complexity of geometry in tensegrity grids. In this section, non-contiguous strut configurations are summarized. The resulting infinitesimal mechanisms at the connecting joints can be avoided by the same principle as that in non-contiguous strut tensegrity configurations (Section 3.2.1) and the discussion is omitted here.

4.4.1 Configurations made of APs and ATPs

The configurations made of APs (Figure 4.9) are the same as those of tensegrity prisms, which can be referred to in Chapter 3. In contrast, prism rotation angle in APs can be smaller than 30° for a triangular AP and 45°

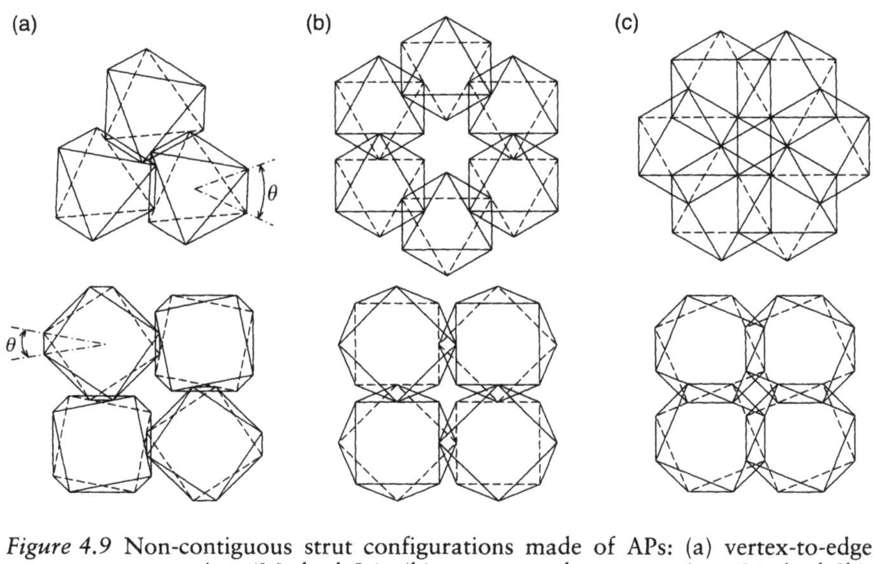

Figure 4.9 Non-contiguous strut configurations made of APs: (a) vertex-to-edge connection (Method Ia); (b) vertex-to-edge connection (Method Ib); (c) edge-to-edge connection (Method II).

for a square AP. So Method Ia, which is not suitable for the square tensegrity prism, is suitable for the square AP. For the other two methods (Methods Ib and II), regular APs are preferred. In Method Ib, additional cables are required to restrain relative rotation of adjacent simplexes.

The configurations made of ATPs are the same as those of APs but are applied in the domical forms with camber. For example, the Swiss Expo 2001 (Pedretti 1998), as the rare application proposal in the field, is based on Method Ia, triangular ATPs.

4.4.2 Configurations made of Ps and TPs

Non-contiguous strut configurations made of Ps can be classified into two types: face-to-face connection (Figure 4.10(a)) and edge-to-face connection

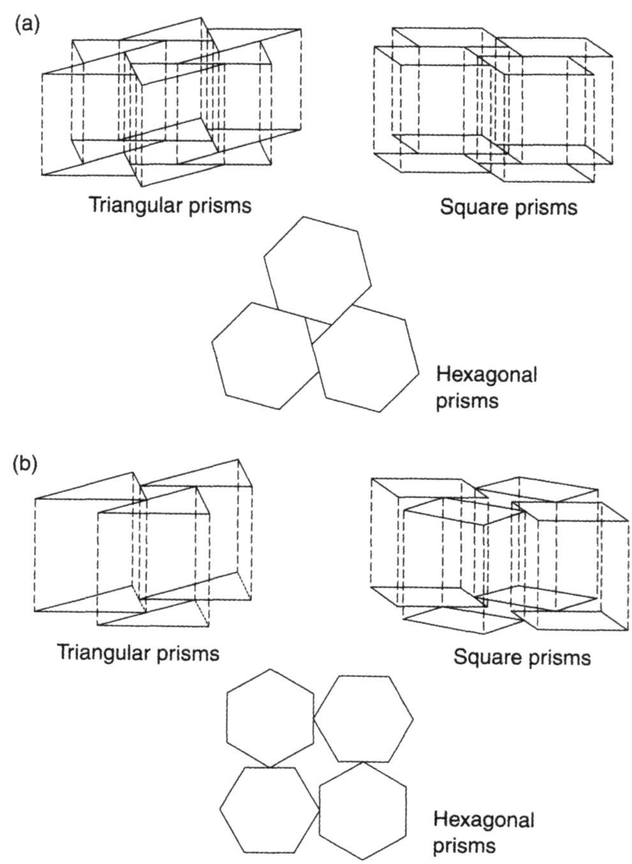

Figure 4.10 Non-contiguous strut configurations made of Ps: (a) Face-to-face connection; (b) Edge-to-face connection.

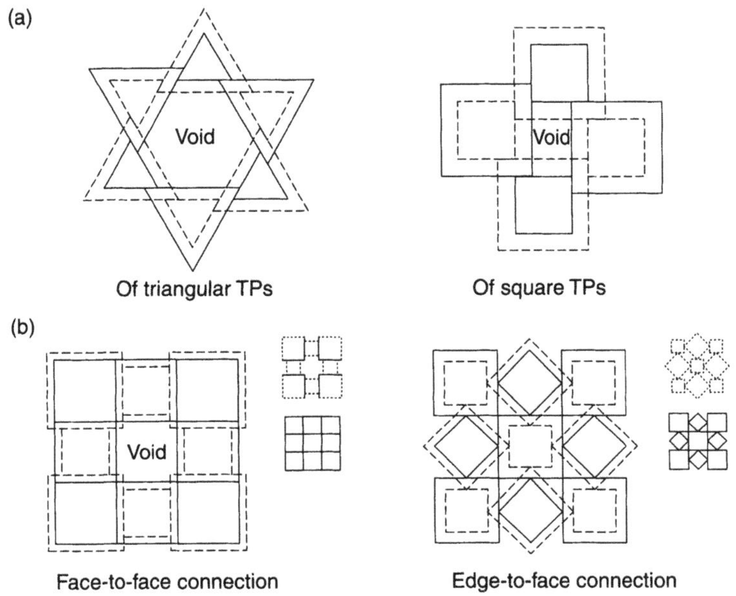

Figure 4.11 Non-contiguous strut configurations made of TPs: (a) face-to-face connection; (b) connections of square TPs (half non-contiguous forms).

(Figure 4.10(b)). In the former, adjacent simplexes share a part of lateral face. In the latter, the vertical edge of one simplex is connected to the lateral face of another simplex. In the edge-to-face connection of square-based simplexes, additional cables are required to restrain relative rotation of adjacent simplexes, which are not shown in the figure.

The configurations made of TPs can be got from the varied domical forms of Ps. But some flat forms are also feasible, for example, the face-to-face connection of triangular or square TPs (Figure 4.11(a)). In addition, some 'half' non-contiguous forms can be derived, in which struts are connected directly just in one layer (Figure 4.11(b)).

4.4.3 *Configurations made of CPs and DPs*

The configurations made of CPs and those of DPs are basically the same. The bases are connected in either a vertex-to-edge way or an edge-to-edge way (Figures 4.12 and 4.13). Bottom cables are needed in CP configurations to form the tensional layer. The resulting CP grids act as free-standing grids when downward load is dominant. In DP configurations, upper and bottom cables are introduced so that two layers of cables can act as the tensional layer alternatively under various loads. The discontinuity of strut groups, as shown in Figure 4.12, present unique appearance.

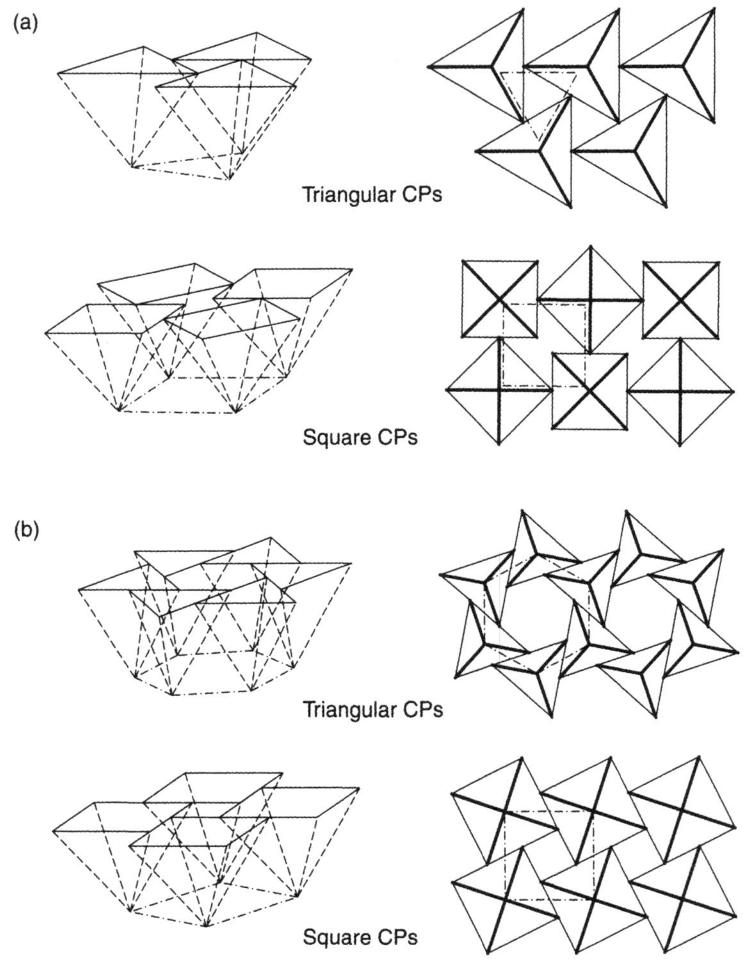

Figure 4.12 Non-contiguous strut configurations made of CPs: (a) vertex-to-edge connection; (b) edge-to-edge connection.

4.5 Cable-strut grids of contiguous strut configurations

Contiguous strut configurations made of cable-strut simplexes are the basic structural configurations for designing lightweight cable-strut grids. The configurations are summarized in this section. In free-standing supporting structures, geometrically rigid forms achieve better stiffness and more even distribution of internal forces and even obviate prestressing process. Most configurations are geometrically rigid themselves. The left with mechanisms are discussed and improved into geometrically rigid ones. Besides, some evolved structural configurations and shell forms are illustrated.

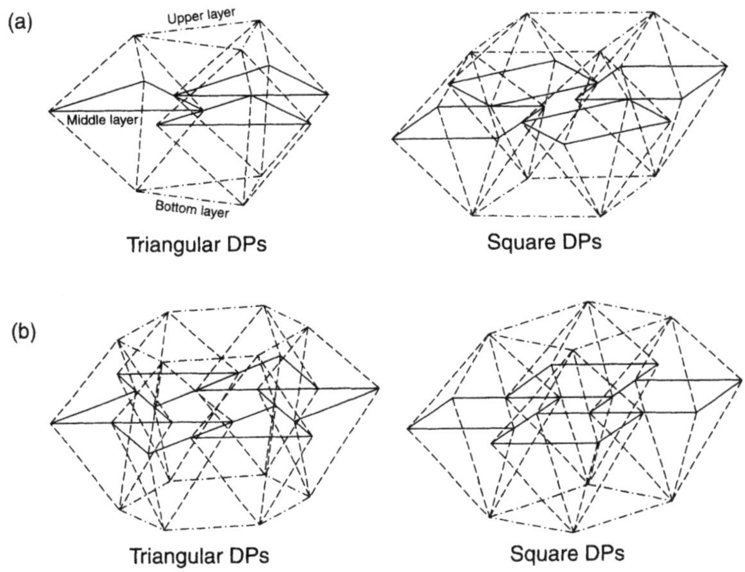

Figure 4.13 Non-contiguous strut configurations made of DPs: (a) vertex-to-edge connection; (b) edge-to-edge connection.

4.5.1 Configurations made of APs and ATPs

All connection methods for the contiguous strut configurations made of tensegrity simplexes can be applied into the corresponding APs and ATPs (Figure 4.14). In flat grids, the vertex-and-edge connection is suitable for ATPs whereas the vertex-to-vertex connection is suitable for APs. Meanwhile, attaching cables are also introduced in each vertex-to-vertex connection like the case in tensegrity configurations, in order to reduce internal forces and increase structural stiffness. The configurations made of AP2s and ATP2s are the same as those of APs and ATPs.

Unlike in the configurations made of tensegrity simplexes, there is in general no chirality in the configurations of the square AP or ATP and regular APs and ATPs are introduced more often to achieve better structural properties. In addition, an additional form can be feasible for the triangular AP in the vertex-to-vertex connection (Figure 4.14(c)), in which each AP should be regular. The connection is better than the vertex-and-edge connection of the triangular ATP (Figure 4.14(a)) in that its bar density is lower. Meanwhile, the form in Figure 4.14(b) can be considered as the honeycombed form of Figure 4.14(c) when each simplex is regular, as is the case in the figure.

The connection of the square AP (Figure 4.14(e)) and that of the triangular AP in Figure 4.14(b) have the drawback that each pair of connected

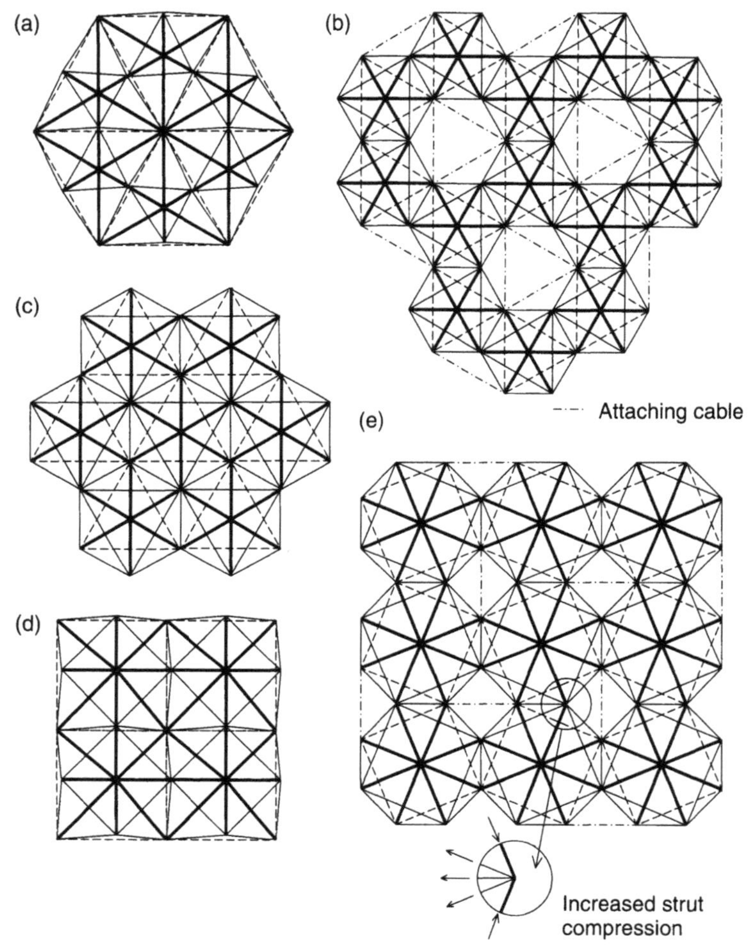

Figure 4.14 Contiguous strut configurations made of APs and ATPs: (a) vertex-and-edge connection of the triangular ATP; (b) vertex-to-vertex connection of the triangular AP (Layout (a)); (c) vertex-to-vertex connection of the triangular AP (Layout (b)); (d) vertex-and-edge connection of the square ATP; (e) vertex-to-vertex connection of the square AP.

Source: Wang and Li (2001); Courtesy: Multi-Science Publishing.

struts from adjacent simplexes is inclined to each other in plan. Therefore, additional force component is required for equilibrium (see also analysis in Section 5.3.1). So the vertex-and-edge connection (Figure 4.14(d)) is better for the square simplexes, whereas the vertex-to-vertex connection in Figure 4.14(c) is more suitable for the triangular.

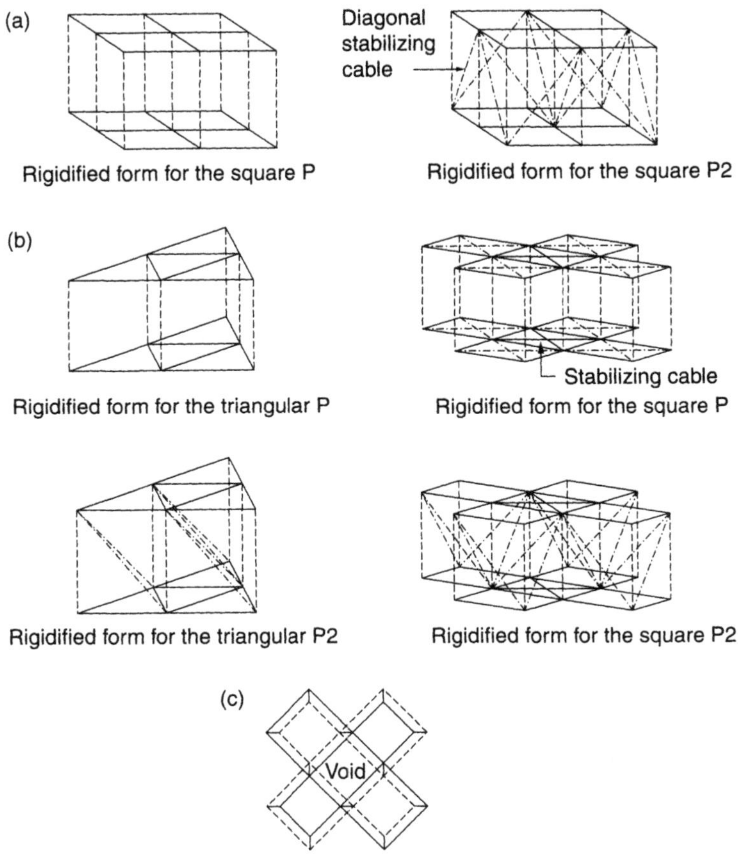

Figure 4.15 Contiguous strut configurations made of P (TP) simplexes: (a) face-to-face connection; (b) edge-to-edge connection; (c) square TP in flat form (edge-to-edge).

4.5.2 *Configurations made of Ps and TPs*

The configurations of prisms can be classified into the face-to-face connection method (Figure 4.15(a)) and the edge-to-edge connection method (Figure 4.15(b)). The face-to-face connection is not suitable for the triangular simplexes due to high bar density, which also leads to tensional bars that are considered 'redundant' under cable-strut concept. In the edge-to-edge connection of the square simplexes, crossing cables are added in both layers to avoid relative rotation although these cables sustain only small forces under load.

In the face-to-face connection, the mechanisms in the base and those of base rotation are avoided by the restraints from adjacent simplexes. Consequently, configurations made of the square P are geometrically rigid. In the connection of the square P2, the left mechanisms of translation are avoided by introducing one diagonal cable in each lateral face (Figure 4.15(a)).

In the edge-to-edge connection, only the mechanism of base rotation can be avoided, but it is enough for the connection of the triangular P to be geometrically rigid. In the connection of the square P, the mechanism in the base can be avoided by introducing crossing cables in the base of the tensional layer (in Figure 4.15(b), both layers). In the connection of the triangular or square P2s, stiffened simplexes have to be used in order to achieve rigidified forms.

All configurations made of P simplexes can be applied to the corresponding TP simplexes to form curved forms. It seems not convenient to form flat TP grids. However, one such special case is presented in Figure 4.15(c), which is the edge-to-edge connection of the square TP of alternate large and small bases.

4.5.3 Configurations made of CPs

With reference to the methods of forming AP and ATP grids, contiguous strut configurations of CPs fall into two categories: edge-to-edge connection (Figure 4.16(a)) and vertex-to-vertex connection (Figure 4.16(b)). In both

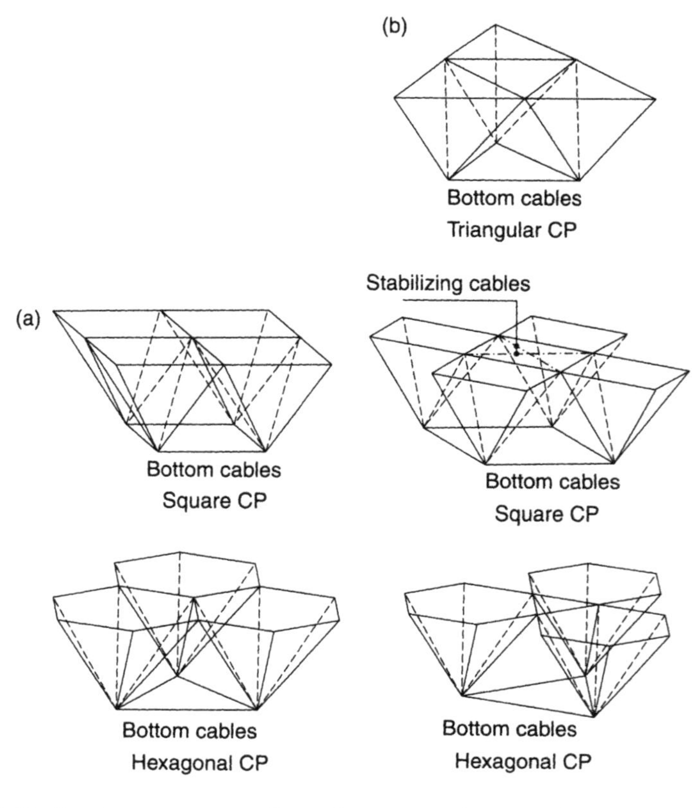

Figure 4.16 Contiguous strut configurations made of CP simplexes: (a) edge-to-edge connection; (b) vertex-to-vertex connection.

connections, bottom connecting cables are added to form the tensional layer.

For the triangular CP, the edge-to-edge connection is not suitable. For the square, stabilizing cables are added among simplexes to avoid relative rotation in the vertex-to-vertex connection. The connections of the hexagonal simplexes are also illustrated for reference. All CP configurations are geometrically rigid.

As the special case of cable-strut grids, the CP grids are not restrictedly free-standing as bottom cables cannot form a compressive layer. But they behave as free-standing forms when the gravity load (including roofing, cladding and self-weight load of the grid) is dominant. Special configurations refer to Section 7.3.3.

4.5.4 Configurations made of RPs and DPs

As far as various RP simplexes are concerned, their configurations can also be composed of two types: vertex-to-vertex connection and edge-to-edge connection. Two layers of cables are introduced in forming RP configurations. For the triangular simplex, only the vertex-to-vertex connection is suitable (Figure 4.17(a)). For the square simplex, both connection methods are suitable (Figure 4.17(b) and (c)), and in the vertex-to-vertex connection relative planar rotation of adjacent simplexes may be avoided by introducing crossing cables in the grid or bars along the edge. Configurations of the hexagonal simplex are presented in Figure 4.17(d) and (e) for reference. All RP configurations are geometrically rigid.

Each RP configuration has its corresponding form in space trusses if we change the reciprocal prism to the pyramid of the same base and restore connecting cables to bars. For example, the RP grid formed by the edge-to-edge connection of the square simplex corresponds to the square-on-square offset space grid (the 'SOS' grid), and the one formed by the vertex-to-vertex connection corresponds to the diagonal-on-square space grid (the 'DOS' grid). Other sub-types of RP grids can be formed by referring to the methods of forming various space trusses. Their composition being compared at the same modular length, RP grids reduce bar density significantly with typical proportion of 3:8 by replacing all inclined and bottom bars with cables and only one vertical strut in each simplex. Proportion of cables in RP grids to bars in space trusses is typically 12:8. But as cables are not subject to buckling and large part of them is of low stress level, the overall structural efficiency is higher than the latter, as proved in Section 5.3.

The configurations made of DP simplexes are the same as those of RP simplexes (Figure 4.18). In the vertex-to-vertex connection of the square DP, the introduction of prestressed crossing cables can stabilize the relative rotation. But introducing a strut is preferred for 'purely' hinged node model. When moment resistance of nodes is considered, stabilizing components may be obviated as discussed in Section 5.3.1.

(a)

(b)

(c)

(d)

(e)

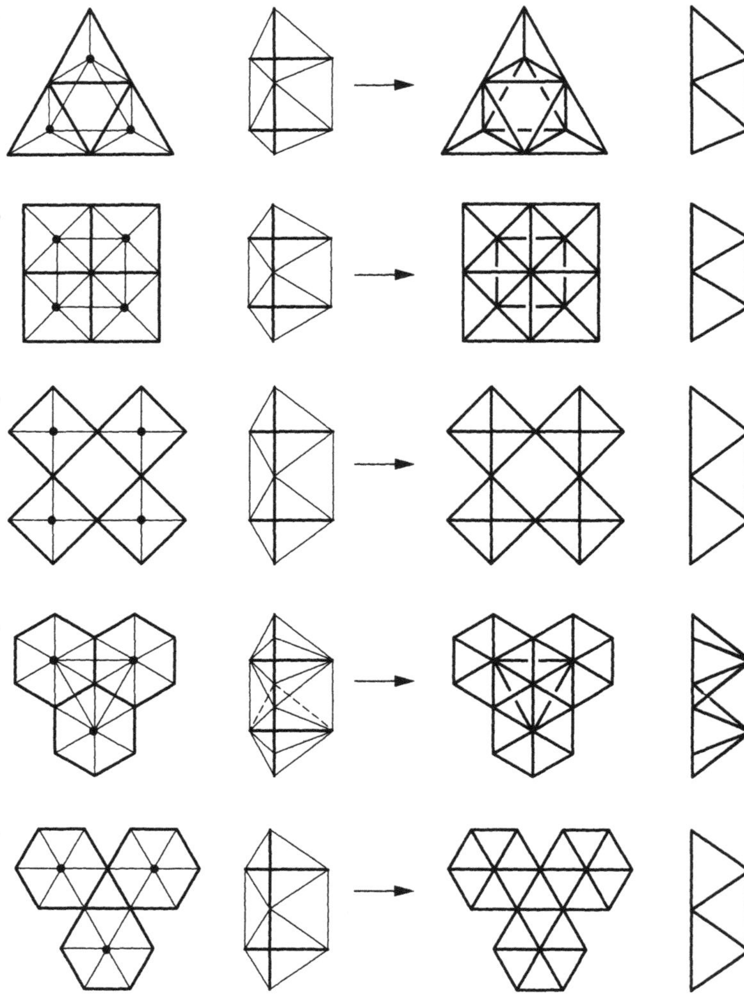

Figure 4.17 Contiguous strut configurations made of RPs: (a) vertex-to-vertex connection of the triangular RP and its corresponding space truss form; (b) edge-to-edge connection of the square RP and its corresponding space truss form; (c) vertex-to-vertex connection of the square RP and its corresponding space truss form; (d) edge-to-edge connection of the hexagonal RP and its corresponding space truss form; (e) vertex-to-vertex connection of the hexagonal RP and its corresponding space truss form.

4.5.5 Summary of basic structural configurations

Among configurations composed of cable-strut simplexes, those made of APs and ATPs are 'special' compared with others, as only one type of connection methods is suitable. In the configurations made of P simplexes, the face-to-face

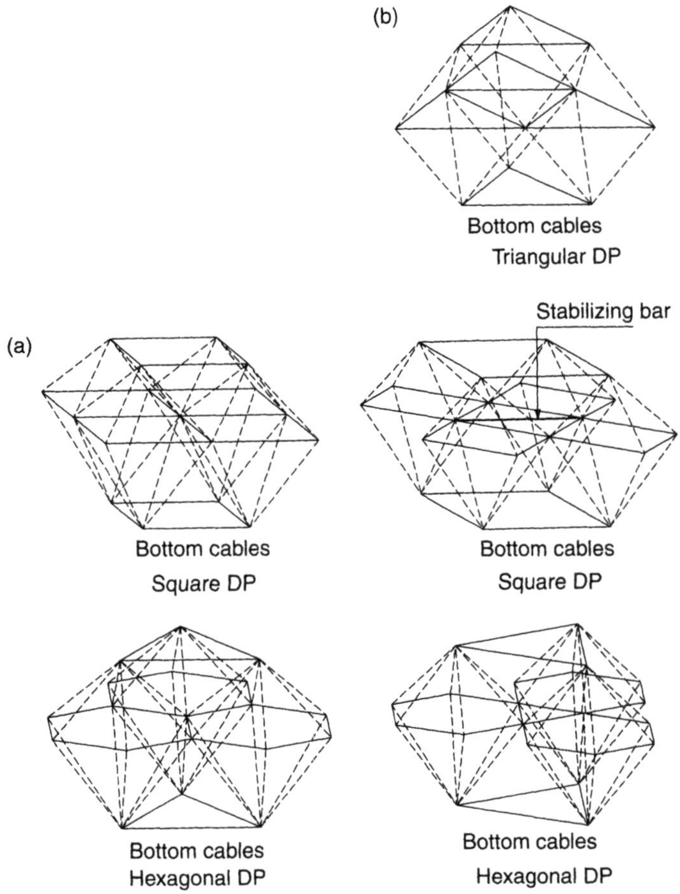

Figure 4.18 Contiguous strut configurations made of DPs (a) edge-to-edge connection (b) vertex-to-vertex connection.

Source: Wang (2002a).

connection method results in more connected joints between each pair of adjacent simplexes than the edge-to-edge connection. The case is the same for the edge-to-edge connection in the configurations made of CPs, RPs or DPs than the vertex-to-vertex connection. For the convenience of narration, the former connection method with more connection joints is named Type 'a', the latter named Type 'b'. The resulting configuration is accordingly called 'P-a' configuration, 'P2-b' configuration, 'CP-b' configuration, and so on, as summarized in Table 4.1.

Configurations in this chapter are obtained from geometrical relations. The Type 'a' of the triangular simplexes is excluded from cable-strut configurations as they contain 'extra' bars that are unavoidably in tension

Table 4.1 Summary of typical cable-strut simplexes and basic structural configurations

Simplexes	*Polyhedra*	*Basic structural configurations (contiguous strut)*	
P, P2	Prisms	Face-to-face (P-a, P2-a)	Edge-to-edge (P-b, P2-b)
CP	Pyramids (crystal-cell pyramids)	Edge-to-edge (CP-a)	Vertex-to-vertex (CP-b)
RP, DP	Riciprocal prisms or di-pyramids	Edge-to-edge (RP-a, DP-a)	Vertex-to-vertex (RP-b, DP-b)
ATP, ATP2	Anti-truncated pyramids	Vertex-and-edge	—
AP, AP2	Anti-prisms	—	Vertex-to-vertex

(actually, Type 'b' can be taken as Type 'a' with openings). Some other forms may also not satisfy cable-strut definition and modification is required, but such cases need to be verified through structural analysis.

4.5.6 Evolved structural configurations

The aforesaid structural configurations are characteristic of co-planar bases as the typical way to form the load-bearing layer. More structural forms can be developed, which are conceptually illustrated as follows. Other forms developed mainly for architectural concern (but may still be structurally efficient) are discussed in Chapter 7.

Grid with openings

In small spans, openings may be designed to reduce material consumption. It is generally suitable for all simplex types. The forms made of triangular simplexes with openings are illustrated in Figure 4.19(a). It is evolved from Type 'b'. In Figure 4.19(b), the forms made of square simplexes with openings can be considered as an evolved form of Type 'a'. When further opening is introduced as checkerboard pattern (Figure 4.19(c)), it is characteristic of Type 'b'.

Mixed base types

Different base types can be mixed in one grid. A simple example, which is just a design technique, is the mixing of triangles and squares to meet the shape of boundary (Figure 4.20(a)). Another example, a mixing of triangles and hexagons with stabilizing cables in plan, among lots of possible forms, is illustrated in Figure 4.20(b).

Mixed simplex types

It is interesting to note that various types of simplexes can be used in one grid. One such example composed of square ATPs and triangular Ps with

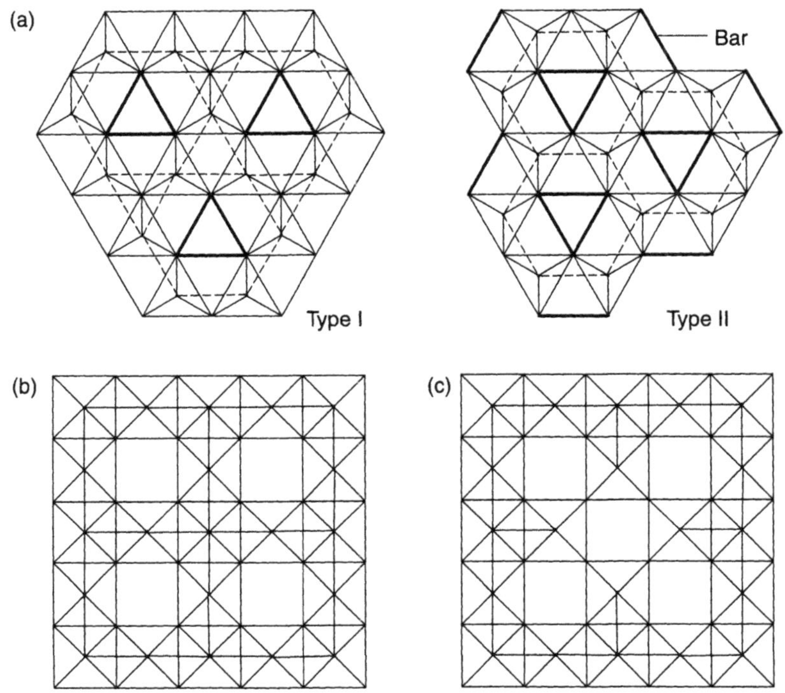

Figure 4.19 Evolved structural configurations: (a) grid of triangular simplexes with openings (from Type b); (b) grid of square simplexes with openings (from Type a); (c) grid of square simplexes of checkerboard pattern (from Type b).

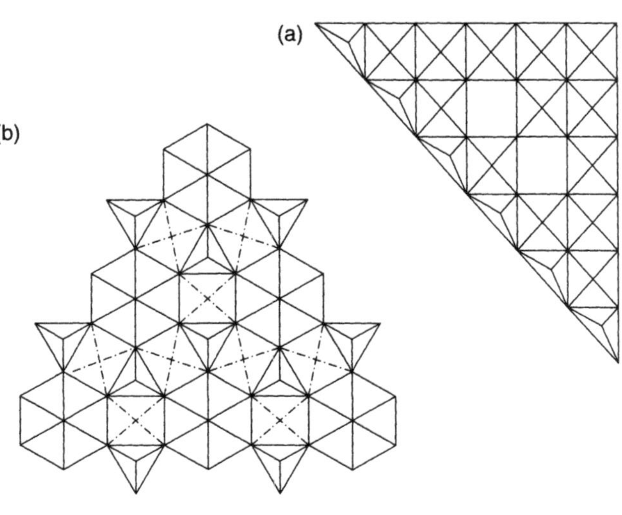

Figure 4.20 Mixed forms of various angle types: (a) mixed form of triangles and squares; (b) mixed form of triangles and hexagons.

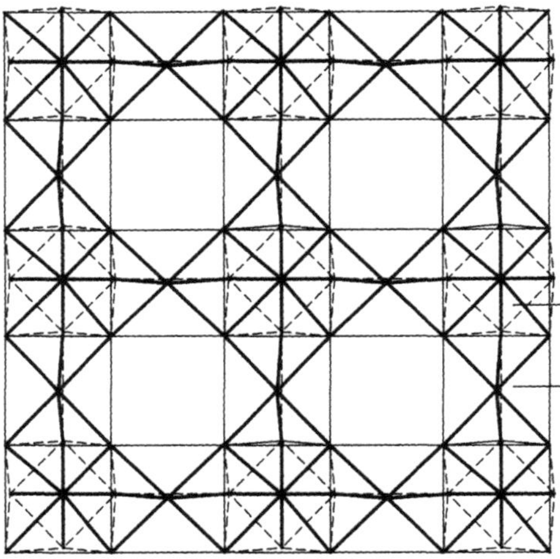

Figure 4.21 Mixed configurations made of square ATPs and triangular Ps.

openings is presented in Figure 4.21. The triangular Ps are 'laid down' with two bases connected to the lateral faces of ATPs. It is for sure that more forms can be designed.

4.6 Cable-strut shells

Latticed shells benefit from space action to lower self-weight. Large deformation is not permissible because joints will deviate from the original positions, destroying space action. Cable-strut grids discussed previously are based on flat forms, which can be developed into low-rise domes. When cable-strut shells, especially cylinders (barrel vaults) are designed in large spans, improvement of structural stiffness may be necessary.

The flat RP or DP grids studied previously are of single-layer bar form, thus the corresponding grids are subject to relatively large deformation. When applied in cylindrical forms, RP-a and DP-b grids are more suitable. The structural stiffness can be improved through introducing bars to replace connecting cables. Therefore, double-layer bar form can be obtained (Figure 4.22(a)), in which all the bottom connecting cables are replaced by bars, and so does triple-layer bar form (Figure 4.22(b)), in which both the upper and bottom connecting cables are replaced by bars. Alternatively, connecting cables along generatrix (in a dome, circular direction) can be kept to form partial double-layer/triple-layer bar forms (Figure 4.22(c) and (d)). Based on the same principle, (partial) double-layer bar form for CP grids can be formed (Figure 4.22(e) and (f)).

In double-layer and triple-layer strut forms cable weight takes very small proportion as cables are just stabilizing elements. Due to the savings in web components, cable-strut shells are lighter than latticed bar shells.

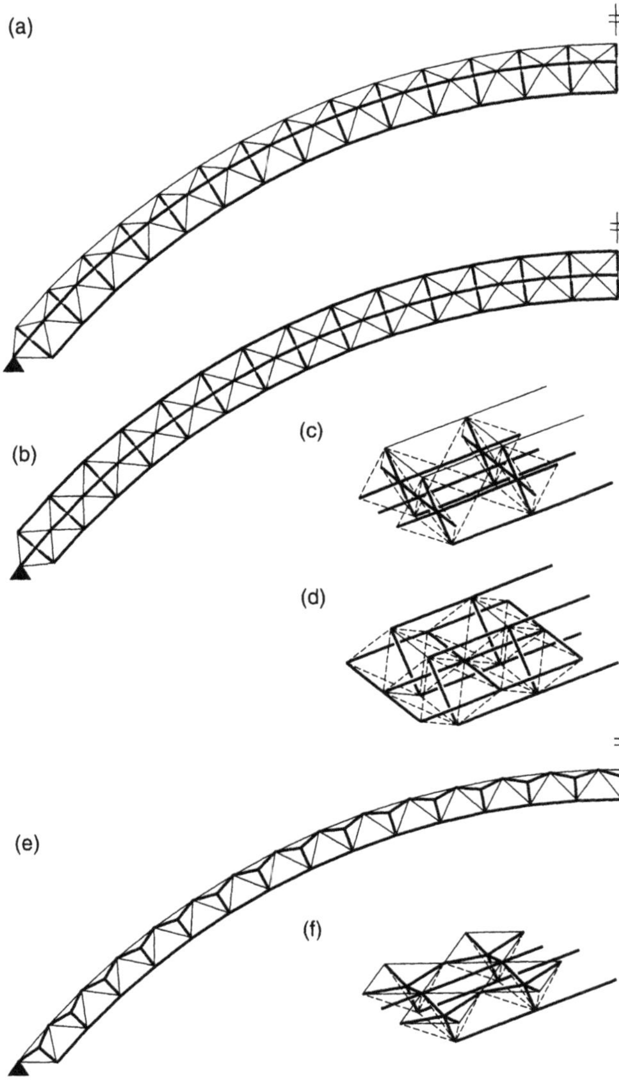

Figure 4.22 Cable-strut shells of multi-layer strut forms: (a) double-layer bar form (half span, for RP or DP grids); (b) triple-layer bar form (half span, for RP or DP grids); (c) partial double-layer bar form (for RP or DP grids); (d) partial triple-layer bar form (for RP or DP grids); (e) double-layer bar form for CP grids (half span); (f) partial double-layer bar form CP grids (half span).

Compared with double-layer latticed barrel vault, test computations at 100 m span show that partial triple-layer RP barrel vault can be 14% lighter due to the savings in web bars. DP barrel vault is still lighter due to the reduction in bar length. CP barrel vault is the lightest but under restriction of span.

In general, in small spans, it is sufficient for RP/DP cylinders (barrel vaults) to design single-layer bar form; in middle spans, (partial) double-layer bar form, including CP cylinders (barrel vaults), may be preferred; whereas in large spans (partial) triple-layer bar form may be required. In domical forms, however, the span range is much increased, as domes are naturally stiffer than cylinders (barrel vaults).

4.7 Summary of cable-strut simplexes and grids

As load-carrying grid structures, cable-strut grids, in the general sense, can be classified into four categories based on simplex types and simplex connections:

- struts not in contact in both simplexes and grids;
- struts not in contact in simplexes but in connection in grids;
- struts in contact in simplexes but not in connection in grids; and
- struts in contact in both simplexes and grids.

The former two types are, respectively, non-contiguous strut tensegrity grids and contiguous strut tensegrity grids. The latter two are new cable-strut grids. The last produces the lightweight type, the resulting structural

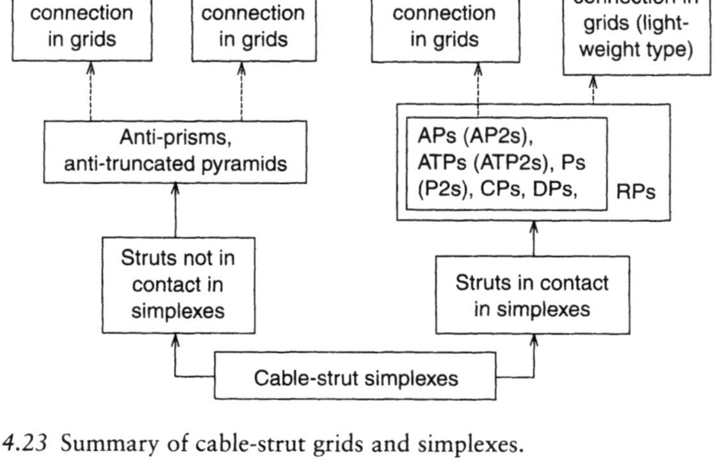

Figure 4.23 Summary of cable-strut grids and simplexes.
Source: Wang and Li (2001); Courtesy: Multi-Science Publishing.

weight of most grids can be lighter than space trusses. The broad sense of cable-strut grids includes tensegrity grids and new cable-strut grids in which struts are allowed in contact in simplexes, whereas the narrow sense includes only the last lightweight type that entirely discards the concept of 'isolation of struts'.

Tensegrity simplexes are anti-prisms and anti-truncated pyramids stabilized by isolated struts from inside and inter-connected cables from outer layer. Cable-strut simplexes are formed from simple polyhedra, including APs and AP2s (formed from anti-prisms), ATPs and ATP2s (formed from anti-truncated pyramids), Ps and P2s (formed from prisms), CPs (formed from pyramids), and RPs and DPs (formed from reciprocal prisms or di-pyramids), etc. Tensegrity or cable-strut simplexes can be applied into non-contiguous strut and contiguous strut configurations. Various forms of cable-strut simplexes and grids are summarized in Figure 4.23. Configurations of lightweight cable-strut grids include Type 'a' and Type 'b' (Table 4.1). In addition, RP, CP and DP grids have multi-layer bar forms, applicable in large span high-rise shells.

In addition to the role in basic structural grids, cable-strut simplexes are basic building blocks for a wide variety of ornamental forms and cable-strut simplexes themselves have varied forms, enriching the application of cable-strut systems. The topic is discussed in Chapter 7. Moreover, a wide variety of irregular polyhedra or higher polyhedra are applicable as building blocks, and grids not based on polyhedra may also be feasible although these themes are not focused in this book, awaiting further investigation.

5 Structural properties and design of lightweight cable-strut grids

In this chapter, the load-transfer properties of lightweight cable-strut grids are explained through truss analysis method based on the simplified planar analogue models. These free-standing modules composed of cables and struts add new content to bar system structural mechanics. Properties discussed include the function and efficiency of various types of grid components in sustaining load response, especially those forming the compressive and tensional layers. Principles of designing cable-strut grids are summarized according to their mechanical properties and are illustrated by design examples. Finally, various aspects of structural properties concerning lightweight cable-strut grids, including weight reduction, stiffness and grid depth are reviewed.

5.1 Truss analysis method on mechanical properties

In practical applications, each grid can be supported and loaded in various ways. Consequently, the distribution of internal forces is somewhat complex. For the reason the properties of the simplified planar analogue models – planar (2D) cable-strut trusses are explained instead to show grid properties conceptually. In this chapter, the terms 'shear force' and 'moment' mean 'shear force on the section as a whole' and 'bending moment on the section as a whole', respectively.

5.1.1 Planar bar truss and tension trusses

Cable-strut grids are both a type of space frames and a type of tension structures. Before the introduction of cable-strut trusses, a planar bar truss (Figure 5.1(a)), representing the load-transfer patterns of space trusses, and tension trusses (Figure 5.1(b), (c)), representing the load-transfer patterns of cable roofs, are presented for reference. Space trusses and cable roofs can be taken as their spatial applications.

Planar bar truss

A planar bar truss is generally called a 'planar truss' in structural mechanics (but as cable-strut trusses studied as follows are also 'planar trusses',

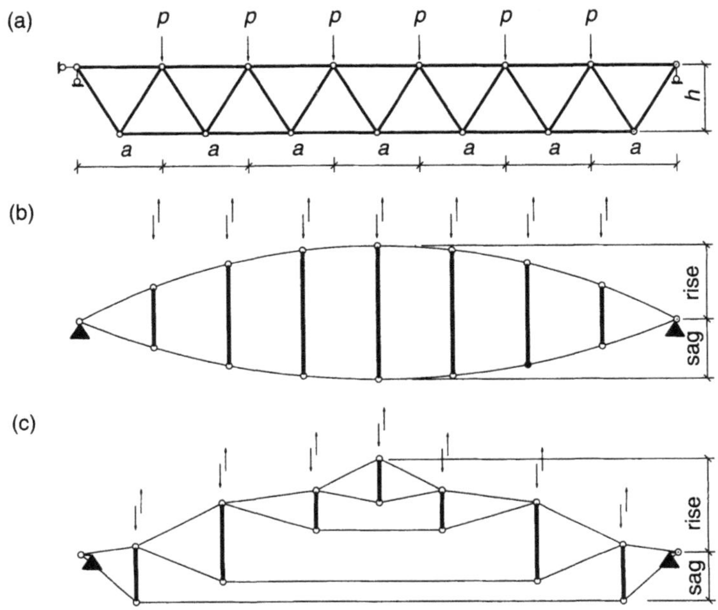

Figure 5.1 Resistant lever arms for the bar truss and tension trusses: (a) planar bar truss; (b) radial cable roof; (c) cable dome (Georgia dome profile).

identification is necessary). Such truss is a simply-supported beam truss (beam-like type) characteristic of large span-to-rise ratio.

Like in a beam, internal forces in a planar bar truss are determined by two types of force equilibrium conditions at each cross-section, namely, shear force equilibrium and moment equilibrium due to external forces. Between the two, shear forces take effect mainly near supports, whereas moments are dominant due to the shallow profile. Inclined chords sustain shear forces. Upper and bottom chords are subjected to compressive and tensional components of internal moments respectively to balance external forces. The functions of these components are the same as in the corresponding space grids. The resistant lever arms for compressive and tensional chords are both the full grid depth, the limit value.

Tension trusses

As catenary-like type, tension trusses are not free-standing and cables are the principal structural components.

In the tension truss for the radial cable roof (bicycle-wheel roof), Figure 5.1(b), top cables slacken under downward load and the resistant lever arm of bottom cables is their sag. Tensions in bottom cables form internal couple with the compression ring beam (in the figure shown as fixed support). Under uplift load, the role of two families of cables is reversed. The relative sags of two families of cables can be adjusted in accordance with the relative magnitude of downward and uplift loads.

The planar tension truss for a cable dome (the shallow profile of Georgia dome, Figure 5.1(c)) shows similar load-transfer pattern to a radial cable roof but is structurally less efficient under downward load. As reported by Hanaor (2002), downward load on internal panels is transmitted via the diagonal cables to vertical struts, which are supported by hoops cables at the bottom, until the vertical struts adjacent to supports. Hence, struts receive enlarged load values with the interaction of edge cables. An additional factor negative to structural efficiency is the near-mechanism geometry of hoop cables in sustaining radial tensions, which cannot be expressed in the truss model. But detailed discussion is out of the scope.

5.1.2 Planar cable-strut trusses

In comparison, planar cable-strut trusses are free-standing beam-like type composed of struts and cables, adding new content to bar system structural mechanics.

The relation of a planar cable-strut truss to a cable-strut grid is similar to that of a planar bar truss to a space truss or a tension truss to a cable roof. A planar cable-strut truss is composed of planar cable-strut modules that are listed in Figure 4.7 with required cables to form the tensional layer. Therefore, there are basically four types of planar cable-strut trusses:

- CP (crystal-cell pyramid) truss (Figure 5.3(a)).
- P (prism) truss (Figure 5.4). The so-called P truss represents not only the dominant properties of P grids, but also AP (anti-prism) grids and ATP (anti-truncated pyramid) grids.
- RP (reciprocal prism) truss (Figure 5.5(a)).
- DP (di-pyramid) truss (Figure 5.6(a)).

For comparison, a tensegrity truss representing the load-transfer pattern of contiguous strut tensegrity grid is also presented here (Figure 5.2), showing the reduced resistant level arms as analysed in Chapter 3. However, it seems

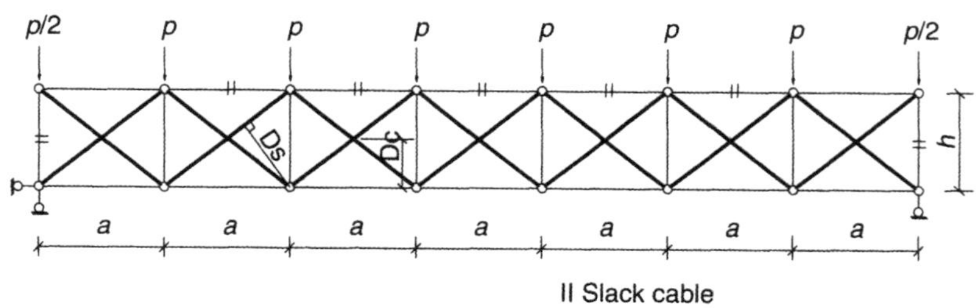

Figure 5.2 A tensegrity truss.

not possible to find a planar truss to represent the pattern of non-contiguous strut types of tensegrity or cable-strut grids, because cables subjected to tension in the compressive layer is due to three-dimensional effect which cannot be modelled in a two-dimensional way.

The relation of cable-strut trusses, except the RP truss, to the corresponding cable-strut grids is straightforward, as the truss itself can be extracted directly from the grid. As a typical example, the relation of the DP truss to two connection methods respectively is illustrated in Figure 5.6(a). Generally, a planar cable-strut truss is formed through projecting a one-way array of square cable-strut simplexes into its symmetrical plane. Examples illustrated are the CP truss (Figure 5.3(a)) and the RP truss (Figure 5.5(a)).

5.1.3 Properties of cable-strut trusses

As each type of planar cable-strut trusses is the simplified form of the corresponding cable-strut grids and is thus characteristic of their load-transfer pattern, the function of each type of components in cable-strut grids is represented by the corresponding components in the cable-strut truss. Properties of cable-strut trusses are analysed as follows, each containing seven modules. No self-stress is introduced for all and various parameters are defined as follows.

- a modular length.
- h overall depth for the CP truss, P truss, tensegrity truss and planar truss; and the distance between the strut layer and the bottom layer (tensional layer under downward load, as always the case studied) in the DP truss and RP truss.
- h_1 the distance from the inner joints to the upper base for the CP truss and P truss.
- h_u the distance between the strut layer and the upper cable layer in the DP truss and RP truss.

The definition of these parameters is the same as those in the corresponding grids. Values typical in the grids are afforded to these parameters for analysis.

In each planar cable-strut module, outer cables are stabilized by inner struts. Under non-prestress state, one cable is slack based on geometrical relations provided that the other cables are in tension and all struts are in compression. The rule is useful in identifying which cable is slack so as to simplify the truss.

CP truss

In the planar CP truss, there are two cable-slackening modes for planar CP modules: upper cable slackening mode and inclined cable slackening mode (Figure 5.3(b)). In the boundary module ABCD where shear force is dominant,

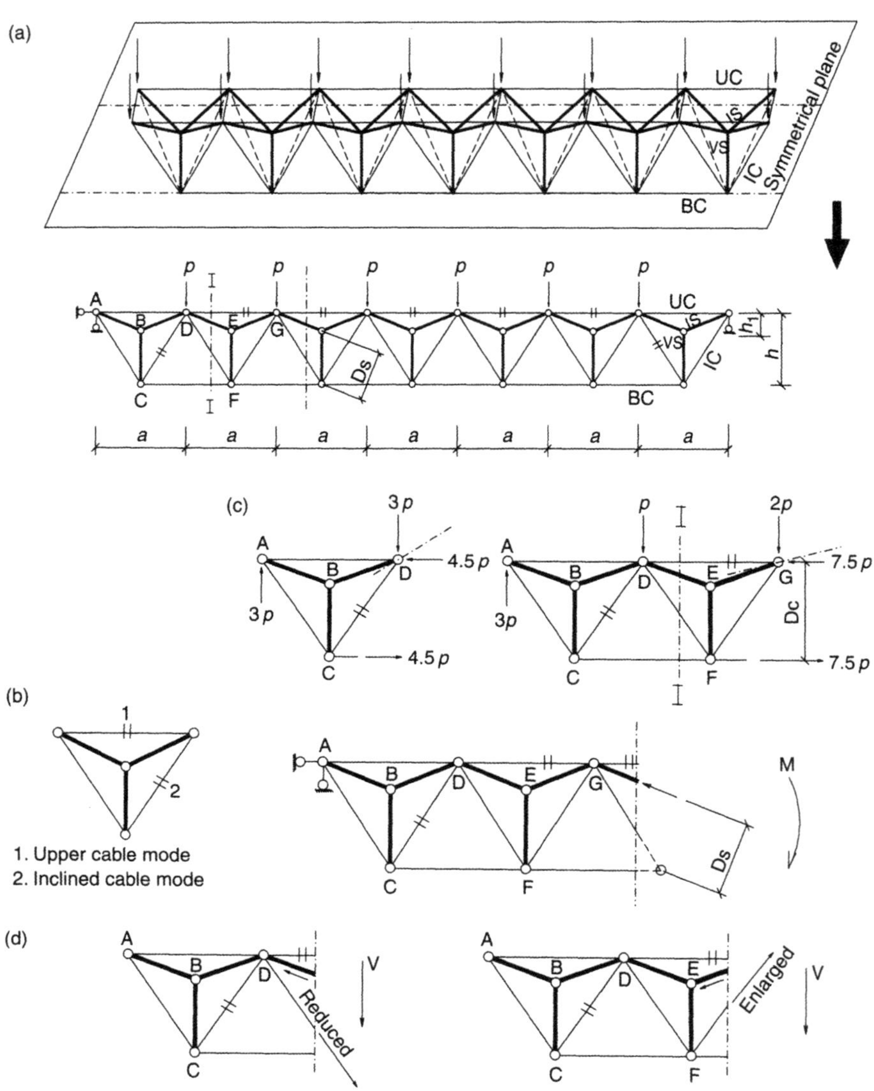

Figure 5.3 Conception and analysis of the CP cable truss: (a) conception of the CP cable truss (II: slack cables); (b) planar CP module; (c) analysis of cable slackening and internal forces ($h_1 = a/6 = h/4$); (d) Shear force reactions in inclined cables.

inclined cable CD slackens as upper cable AD and inclined cable AC are tensional so as to satisfy equilibrium condition at node D (Figure 5.3(c)). Whereas in the inner module DEFG, inclined cable FG is in tension because at node G, the shear force is much smaller than the compressive component of the moment. Meanwhile, inclined cable DF is in tension in order to equilibrate the shear force at the cross-section I-I (Figure 5.3(c)). Hence, upper cable DG slackens. By repeating the procedure, all upper cables in inner modules slacken. Therefore, the upper cable slackening mode is dominant.

The CP truss becomes both statically and kinematically determined after slack cables are removed. The role of the vertical strut is always to brace the module. The tensions in all bottom cables can be determined directly by the moment equilibrium at the section. The corresponding resistant lever arm (Dc) is the overall depth, h (Figure 5.3(c)). But at getting the compressions in inclined struts, those in inner modules and in boundary modules are different. In inner modules, the compressions can be obtained by the equilibrium of sectional moments. The corresponding resistant lever arm (Ds) is reduced owing to bar inclination, so smaller h_1 is preferred (but it cannot be too small so as to avoid instability). In each boundary module, the compressions in inclined struts are obtained by the equilibrium of sectional shear forces so larger h_1 may reduce the compressions. Based on the analysis, it is optimal to design different h_1, that is, smaller at the inner modules and larger at the boundary.

The role of inclined cables is to sustain shear forces. It is interesting to find that due to inclined struts providing vertical force components, forces in the inclined cables lying in the near-support side are reduced whereas those on the other side are increased (Figure 5.3(d)).

P truss

The properties of the P truss (Figure 5.4) are similar to those of the CP truss. Based on the same procedure of analysis, the P truss can also be simplified

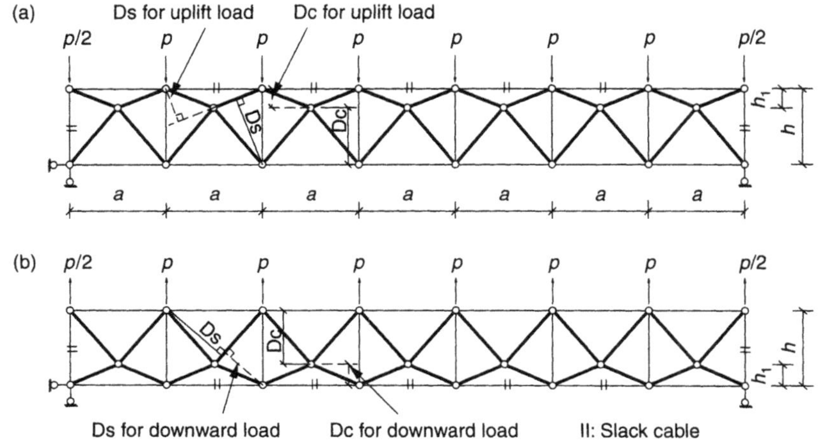

Figure 5.4 A P truss: (a) when downward load is dominant; (b) when uplift load is dominant.

as statically and kinematically determined form by removing slack cables and the internal forces can be determined accordingly.

The geometry of the P truss when the downward load is dominant refers to Figure 5.4(a). Under downward load, slack cables are upper cables in inner modules and two boundary vertical cables connected to the support (it follows the same case as in the tensegrity truss in Figure 5.3). Upper inclined struts form the compressive layer and bottom cables form the tensional layer. The resistant lever arm for bottom cables in inner modules (Dc) is h–h_1, for upper inclined struts in inner modules (Ds) is larger, as marked in Figure 5.4(a). Both resistant lever arms are reduced by bar inclination, so it is desirable to design smaller h_1 in inner modules when downward load is dominant. The lower inclined struts sustain shear forces with reduced values in the near-support side and enlarged value on the other side. When secondary uplift load is applied, upper cables and lower inclined struts form the tensional and compressive layers with smaller resistant lever arm (Figure 5.4(a)), respectively.

When uplift load is dominant, the geometry is reversed, as shown in Figure 5.4(b). Therefore, the resistant lever arms in the P truss can be balanced, and it is an advantage over the tensegrity truss (Figure 5.2). Of course, the advantage vanishes when the uplift load and the downward load are equivalent, if really happened. In Figure 5.4(b), due to tensional support reaction, the connected bottom cable slackens, in place of the connected vertical cable. In addition, struts in tension may be unavoidable if h_1 in boundary modules is smaller than $h/2$.

RP truss and DP truss

In the RP cable truss, under downward load, the bottom continuous cables form the tensional layer, and all the upper cables slacken. Under uplift load only, the upper cables form the tensional layer, and all bottom cables slacken. The horizontal struts always form the compressive layer, and the vertical struts stabilize truss modules. The property is generally similar to the tension truss (Figure 5.1(b)) if the strut layer is replaced by a compressive ring.

Based on the same procedure of slack cable analysis, it is easily seen that the RP cable truss (Figure 5.5(a)) is also statically determinate after slack cables are taken away. Under downward load, inclined cables CD, GF, IJ and the symmetrical ones slacken. Upper inclined cables BD, EG, HJ and the symmetrical ones are subjected to shear forces. The resulting tensions in these inclined cables increase the compressions or equivalently, reducing the resistant lever arm in the corresponding horizontal struts, as shown in Figure 5.5(b) (actually in RP grids, such tension is also sustained by the orthogonal strut in the base that is reduced in the RP truss model). The reduction in lever arm is relatively large near the boundary and is negligible at the middle of span following the decrease of the overall shear forces. In contrast, the resistant lever arm for bottom cables is h.

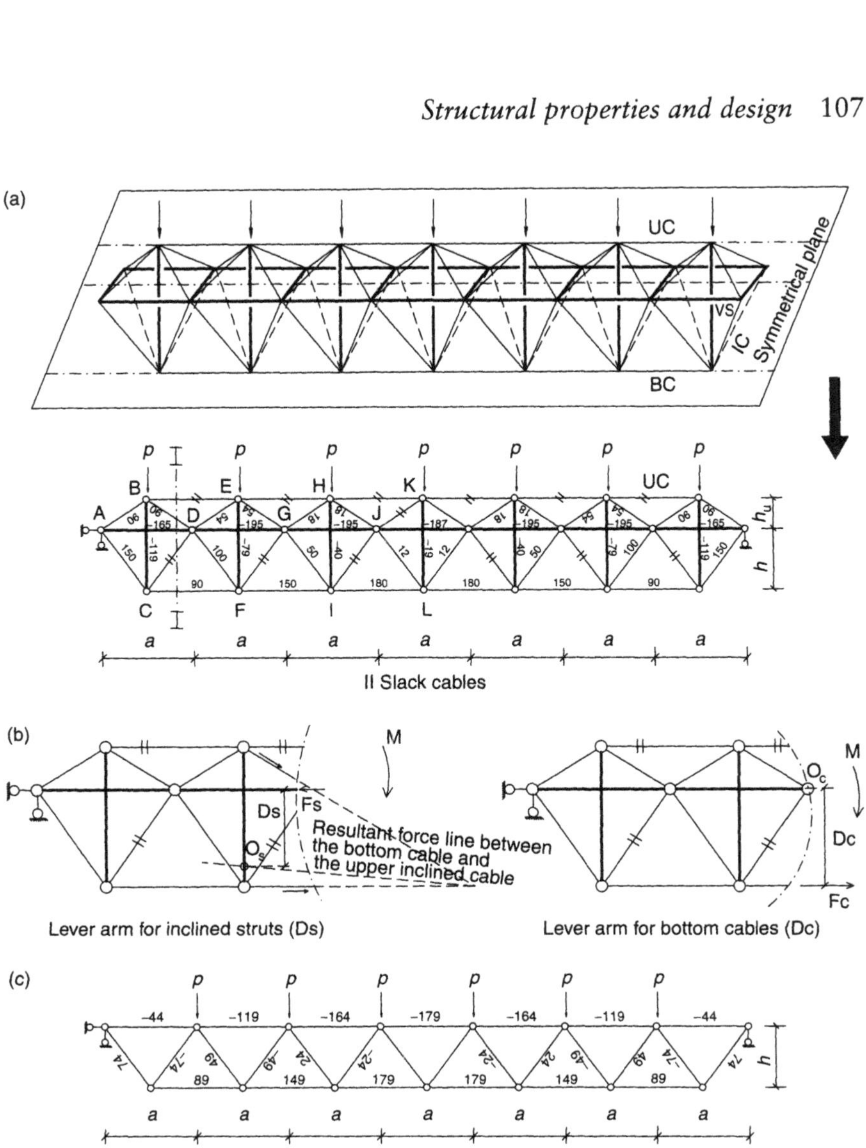

Figure 5.5 Conception and analysis of the RP cable truss: (a) conception of the RP cable truss and internal forces under given parameters ($a = 1$, $h = 2/3$, $h_u = 1/3$, $p = 20$); (b) resistant lever arms for the RP truss; (c) internal forces of a planar truss ($a = 1$, $h = 2/3$, $p = 20$) for comparision.

The internal forces of the RP truss are compared with the planar truss (Figure 5.5(b)) based on the given parameters in the figure for reference. The compressions in the compressive layer of the RP truss distribute evenly at higher values with the maximum value approximate to that in the planar truss. Two trusses are identical in tensions in the bottom layer. So two trusses are generally similar in properties.

The functions of various components in the DP cable truss (Figure 5.6(c)) are identical to those in the RP truss, and the resistant lever arms are also equivalent.

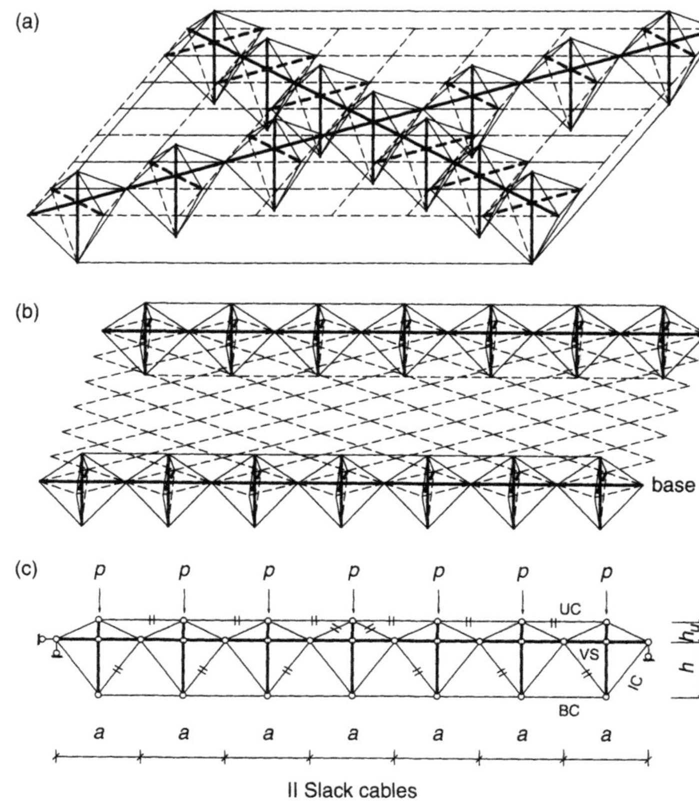

Figure 5.6 Conception of the DP truss: (a) planar DP truses contained in the configuration Type 'a'; (b) planar DP trusses contained in the configuration Type 'b'; (c) a DP truss.

5.1.4 General properties of cable-strut grids

A cable-strut truss represents the load-transfer pattern of the corresponding cable-strut grids. All of them are beam trusses, containing a compressive layer formed by a continuous line of struts and a tensional cable layer, and all struts are in compression. Like the case in the tension truss (Figure 5.1(b)), upper and bottom cables form the tensional layer alternatively under uplift and downward loads, respectively, except the CP truss. It shows the spirit in cable-strut grids, which is proved in design examples in Section 5.3. Optimal parameter design can thus be obtained and adjusting internal forces, when necessary, can have the guide. General properties of each type of cable-strut grids are obtained as follows.

In CP grids, most upper cables slacken under downward load. Some upper cables that do not slacken lie in boundary modules. Inclined struts are dominantly subjected to the compressive component of internal moments with reduced level arms due to bar inclination. Each vertical strut stabilizes inclined struts in the simplex and inclined cables balance shear forces. Bottom cables sustain tensional components of internal moments with the lever arm of h. Designing smaller h_1 in inner modules can increase resistant lever arm for inclined struts that form the compressive layer, whereas designing larger h_1 in boundary modules is preferred.

The properties of P grids, AP grids and ATP grids are similar to those of CP grids. The dominant properties are also characterized by inner modules in which moments dominate the design. Under downward load, upper inclined struts form the compressive layer, and bottom cables form the tensional layer. The resistant lever arm for bottom cables is h–h_1, and for struts is larger. Designing smaller h_1 in inner modules can improve resistance lever arms for both and larger h_1 in boundary modules is preferred. Under uplift load, upper cables and lower inclined struts form the internal resistant couple instead. The corresponding resistant lever arm and design parameters can be obtained accordingly. Under both load cases, balancing of two groups of lever arms shall be considered in determining design parameters.

In RP grids and DP grids, the role of the vertical struts is to stabilize the inclined cables that are subjected to shear forces. Under downward load, bottom cables sustain the tensional components of internal moments with the lever arm of h. Horizontal struts form the compressive layer and the compressions are enlarged by the tensions in the inclined cables. Larger h_u decreases the compressions in the horizontal struts but in RP grids lengthens the vertical struts. Because the effect of bending moments is dominant, the lever arm of the horizontal struts is approximate to h. Under uplift load, upper cables and horizontal struts form the internal couple. The relative value of h_u and h can be adjusted in accordance with the relative magnitude of downward and uplift loads.

5.2 Principles of designing lightweight cable-strut grids

Based on the discussion in the previous section, various cable-strut grids can be summarized into two groups. RP grids and DP grids belong to one group, whereas CP grids, P grids, AP grids and ATP grids belong to the other. In this section, design principles of various lightweight cable-strut grids are presented according to their load-transfer patterns and individual properties. For the convenience of narration, the discussion of design parameters is given representatively for square-based simplexes. Grids made of triangular simplexes are presented in Appendix C for reference.

5.2.1 *General*

In cable-strut grids, components for designing struts are in general round pipes. At larger spans, employing tubes of larger outer diameter (D) and thinner wall thickness (t) are preferred to reduce strut weight. Components for designing cables are tension rods, steel strands, steel wires, etc. In conventional spans, when roof material requires high stiffness, tension rods can be used. In other cases, higher strength material is suitable, resulting in larger deflection which can be offset by camber or slope design.

The joint design in cable-strut grids can be simple as bars are subjected only to compression. But joint design may affect modular length. The topic is discussed in Section 6.2 but is not considered here. In lightweight cable-strut grids, prestress level is in general very low and can be neglected. Most often, the so-called 'prestress' is only used to tighten cables.

In actual application, cable-strut grids may be applied in various supporting conditions including edge-supported, corner-supported, one-way spanning, and even cantilever cases. Conceptually, a grid containing struts diagonal to the edge in the compressive layer is less advantageous in one-way spanning, including the RP-b grid, the DP-a grid, and the CP-a grid, etc. In this section, only edge-supported case is discussed, and the spirit can be extended to other cases.

5.2.2 *RP grids*

When downward load is dominant, parameters for designing RP grids are h_u (the height of the upper part), h (the height of the lower part, normally, the main part), and a (the modular length). As the load-transfer pattern of the RP grids is similar to the corresponding forms of space trusses as a whole, their design parameters can refer to the latter. That is to say, h and a can be the same as those in designing space trusses of the same layouts. h_u is related to h. Designing larger h_u can reduce the internal forces but may increase the weight. The optimal ratio of h_u to h is roughly 0.3, which is got from a number of test designs. When upward load is dominant, the upper part becomes the main part.

Between two types of RP grids, if their modular lengths are designed the same, the base length of each simplex in the RP-b grid is 71% that of the RP-a grid and their self-weight is equivalent. The RP-b grid can be lighter than the RP-a grid if larger modular length is designed. But the modular length is not the larger the better because of bar buckling property. The proportion of two modular lengths shall be clearly smaller than $\sqrt{2}:1$, at which base lengths are the same for two grids. The optimal value is the middle, about 1.2 : 1, and it is proved by test designs.

Moreover, balancing of resistant lever arms (h_u/h) is preferred when a RP grid is subjected to both downward and uplift load cases. For example, when roof dead load = 30kg/m^2, self-weight load = 20kg/m^2, live load = 50kg/m^2, suction load = 70kg/m^2, the resulting design load for the combined downward load case based on Chinese code is

$$30 \times 1.2 + 20 \times 1.2 + 50 \times 1.4 = 130\text{kg/m}^2 \tag{5.1}$$

The design load for the combined uplift load case is

$$70 \times 1.4 - 30 - 20 = 48\text{kg/m}^2 \tag{5.2}$$

Therefore,

$$\frac{h_u}{h} = \frac{48}{130} = \frac{1}{3} \sim \frac{2}{5} \tag{5.3}$$

Normally, the less critical load case (here, the combined uplift load) between the two does not affect weight much as the increase of weight is mainly due to one layer of cables (here, upper cables under the combined uplift load case).

Sample RP grids are given in Figure 5.7 under downward load in 30m span. In the figure, a is 3m for the RP-a grid, 3.75m for the RP-b grid respectively, and h_u, h are 0.75m, 2.5m in both cases.

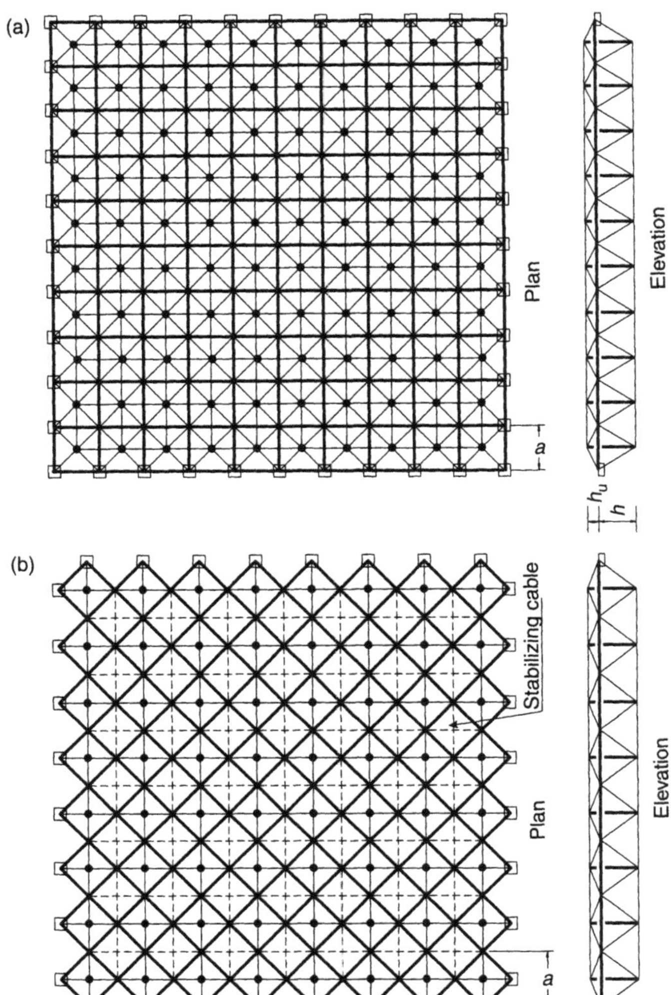

Figure 5.7 Sample RP grids (30m span): (a) RP-a grid; (b) RP-b grid.

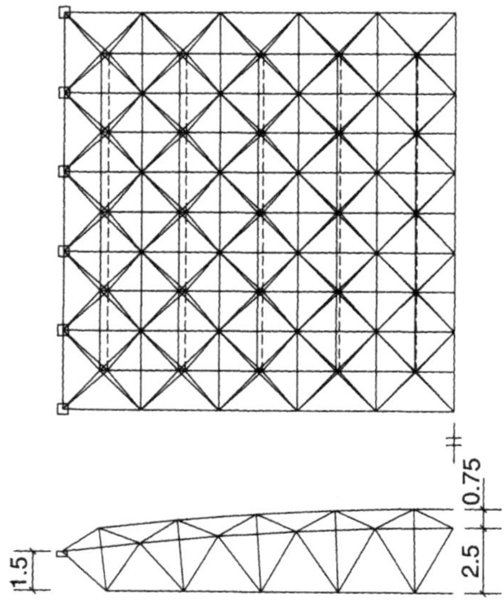

Figure 5.8 Illustration of camber design for the RP grids in 30m span.

In RP grids and other cable-strut grids, camber design can be introduced to consider the water drainage problem and the bottom layer can be designed flat to make even the distribution of internal forces (Figure 5.8). The design can reduce bar weight and deflection in a certain degree.

5.2.3 DP grids

The properties of DP grids are similar to those of RP grids. Sample grids, when downward load is dominant, are given in Figure 5.9 in 30m span. h can refer to the design in RP grids, although larger h may reduce more weight. h_u is affected by joint types. When the joint is of solid ball joint type, for example, h_u is preferred to be larger so as to avoid small included angle among components. Here, h_u/h is recommended to be 0.6. When hollow ball joint or plate-type joint is designed, h_u/h can be smaller.

In the DP-a grid, the modular length a can be larger as strut lengths are short. It can be the same as that in the RP-b grid. The modular length in the DP-b grid is preferred to be the same as that in the DP-a grid. That is because although its horizontal struts are shorter, the forces in these struts are larger due to larger modular length in the strut layer (Figure 5.9). The modular length in the strut layer of the DP-b grid is 3.75m, and that of the DP-a grid is $3.75/\sqrt{2}$m.

Note that inner base cables in the DP-a grid can be removed due to the restraint from adjacent simplexes, as shown in Figure 5.9(a).

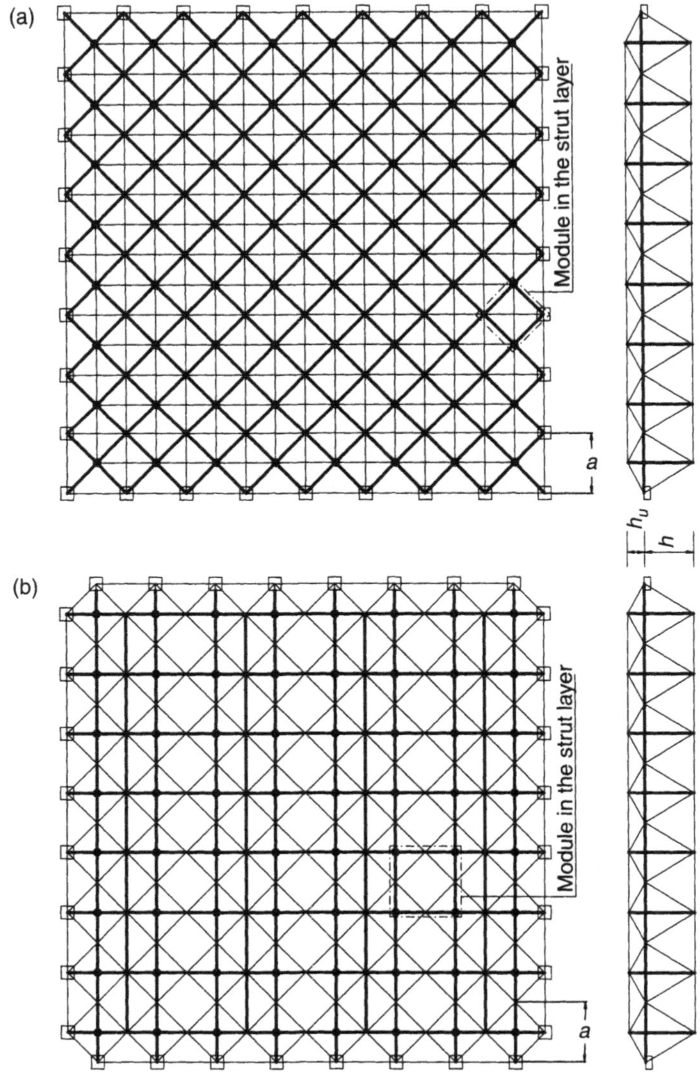

Figure 5.9 Sample DP grids (30m span): (a) DP-a grid; (b) DP-b grid.

5.2.4 CP grids

In CP grids, design parameters are modular length (a), grid depth (h), and the depth from the inner joint to the upper layer (h_1). The majority of the upper cables in CP grids slacken except those at the boundary modules because of the effect of the global shear force. Optimal h_1 is smaller at inner modules (but cannot be excessively small so as to avoid 'snap-through'

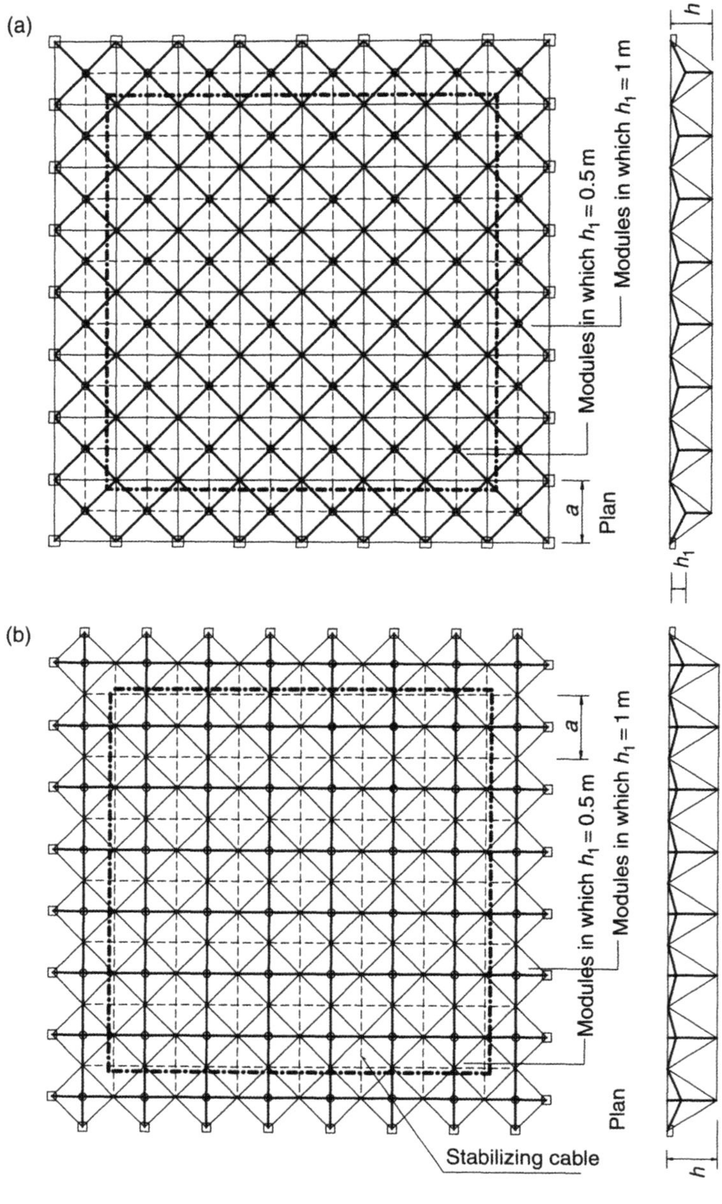

Figure 5.10 Sample CP grids (30m span): (a) CP-a grid; (b) CP-b grid.

buckling) to increase the resistant lever arm for the inclined struts that form the compressive layer (roughly h–h_1), but optimal h_1 is larger at boundary modules. When it is permissible, h can be designed to be relatively larger for weight reduction. The design of a can refer to that in the DP grids.

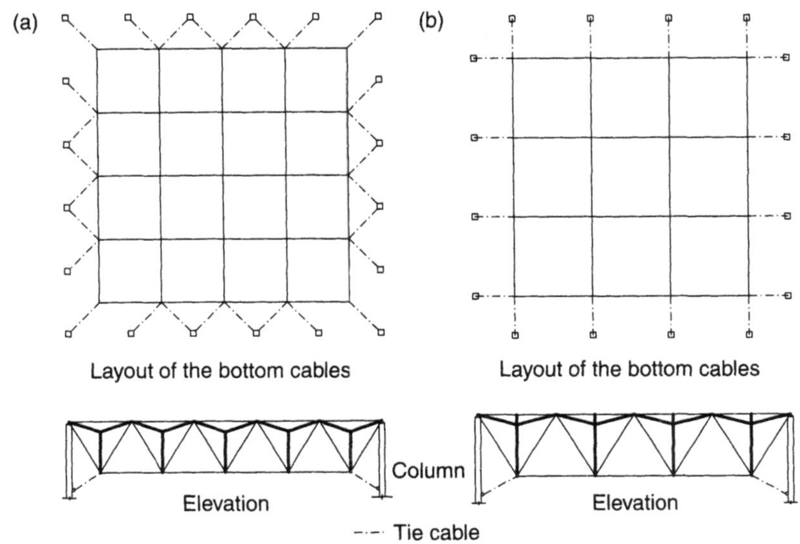

Figure 5.11 Stabilized forms of the CP grids: (a) CP-a grid; (b) CP-b grid.

The design examples for CP grids are given in Figure 5.10. In the figure, h_1 is 0.5m in inner modules, and 1m in boundary modules; *a*, *h* are 3.75m, 2.5m for the CP-a grid, and 3.75m, 3m for the CP-b grid, respectively. *h* in the CP-b grid is designed to be larger in order to reduce compressions as the modular length in the strut layer is relatively large.

In the CP grids, the downward load is in general required to be dominant. They are suitable in the load case that dead load can offset the effect of uplift load. However, when the design uplift load is not much larger than the downward load, the bottom layer may be attached to lateral supports by cables. Sample layouts are illustrated in Figure 5.11.

5.2.5 ATP grids and AP grids

In general, the mechanical properties of the AP and ATP grids are similar to those of contiguous strut tensegrity grids in that in the compressive layer, most cables always slacken, and some subjected to small tensions mainly lie in boundary modules. Therefore, inclined struts connected to the compressive layer sustain the compressive components of internal moments. The difference is that the resistant lever arms for these struts and especially for the bottom cables in the AP and ATP grids can be improved through optimization.

When the ATP grid is subject to dominant downward load case, it is advantageous to place the larger base of each ATP in the upper layer and to design smaller h_1 (Figure 5.12(a)). The optimal design can balance the lengths of struts in the upper and bottom parts in addition to the improvement

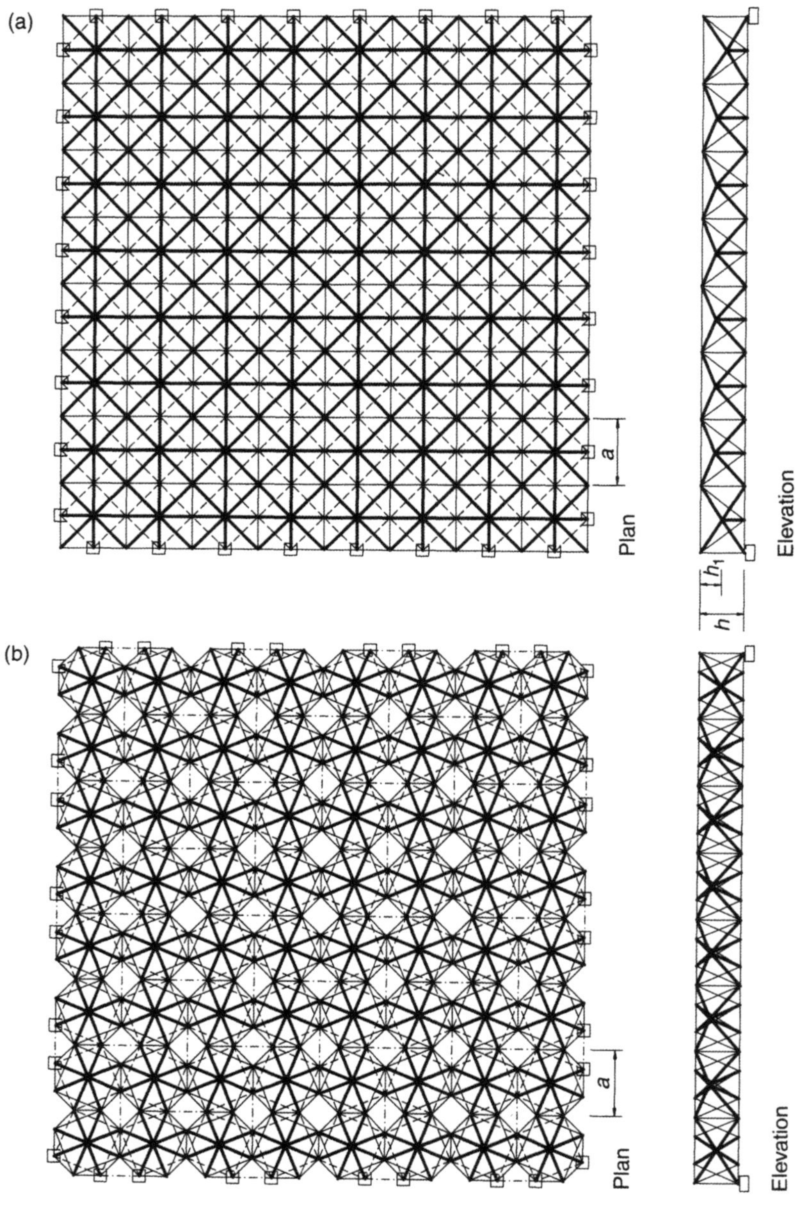

Figure 5.12 Sample ATP grid and AP grid (30m span): (a) ATP grid; (b) AP grid.

in resistant lever arms. The choice of h_1 can refer to the relative design values of downward load and uplift load after combined and the design in the CP-a grid for distribution. In addition to h_1, other design parameters in the ATP grid include module length (a) and overall depth (h) can also refer to those in the CP-a grid.

In the design example shown in Figure 5.12(a) under downward load in 30m span, a is 3.75m, h is 2.5m, and h_1 is the same as that in Figure 5.9(a). When the uplift load is dominant, the grid is placed upside down, and the same design techniques can be applied accordingly.

The design parameters for the AP grid can refer to the ATP grid. The sample grid is given in Figure 5.12(b). However, the grid made of square simplexes is less advantageous as proved by the design examples in Section 5.3. In this case, its prism rotation angle can be adjusted to be smaller or it can be replaced by a P-b grid in application.

The principles of designing ATP2 and AP2 grids are similar and meanwhile, they may also refer to the design of P2 grids in the following section.

5.2.6 *P grids*

In general, between two types of configurations forming cable-strut grids, Type 'a' produces more and longer bars compared with Type 'b'. Therefore, Type 'a' is generally heavier. In the P-a grid, there are high proportion of bars in tension in two-way spanning owing to the interaction of upper and bottom inclined bars, and thus it does not meet the spirit of cable-strut forms. Therefore, openings shall be introduced (Figure 5.13). The case is

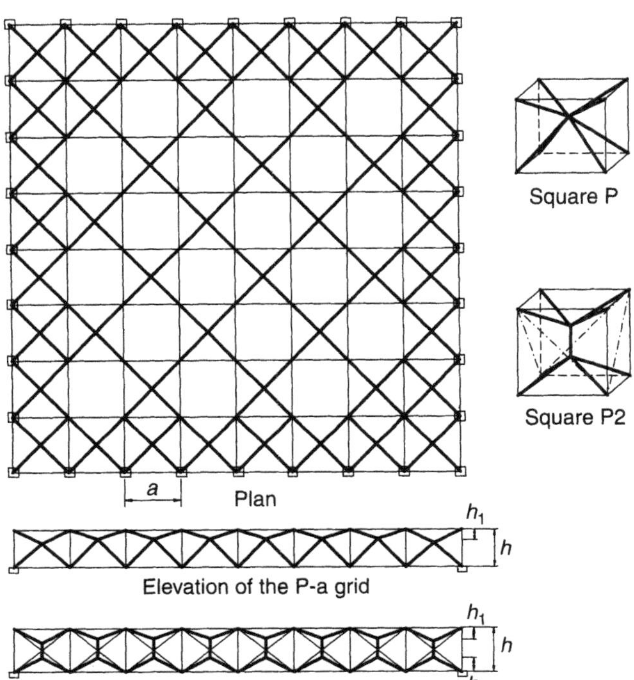

Figure 5.13 Sample P-a and P2-a grids (30m span).
Source: Wang and Li (2003a).

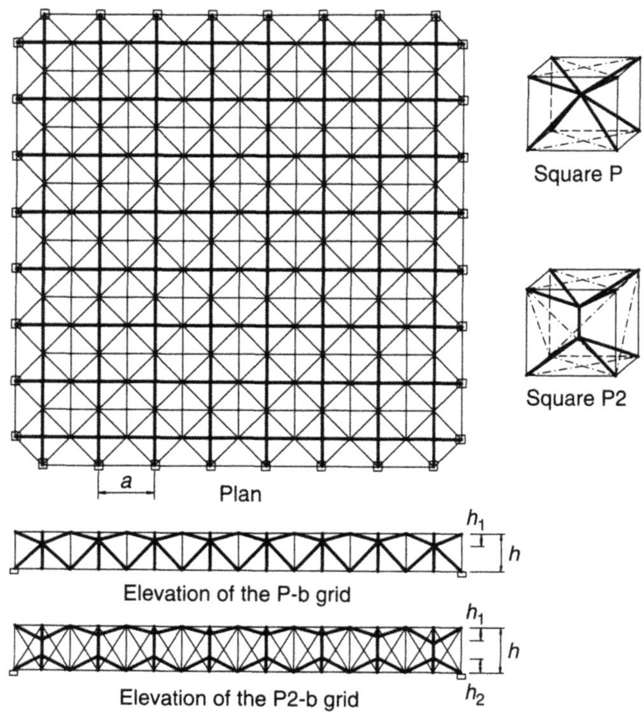

Figure 5.14 Sample P-b and P2-b grids (30m span).

Source: Wang and Li (2003a).

the same for the P2-a grid, in which rigidified simplexes are required to achieve geometrical rigidity.

Sample grids designed under downward load for P grids are shown in Figures 5.13 and 5.14, design parameters including h_1, h and a, which can refer to the design in the ATP grid. Design parameters in P2 grids are h_1, h, a, and h_2 (the distance of the lower inner joint to the bottom base), and optimal h is larger than that in P grids and h_2 is preferred to be larger so as to reduce forces in lower inclined struts.

5.3 Design examples in conventional spans

In this section, various lightweight cable-strut grids are studied in conventional spans (here, 30m). Layouts of these grids are given in Section 5.2. Note that in these contiguous strut forms, the effect of geometrical stiffness is not significant and computation process is stable. Therefore, the design and analysis is easy as it can be solved with sufficient accuracy by lots of commercial software that can model tendons.

For the simplicity of presentation, all grids are studied under downward load only. No prestress is introduced and basic conditions, including loads, span and supporting conditions, are specified to be the same as in Chapter 3. As forces are in general smaller for cable-strut grids, the smallest cross-section for cables is 0.25cm^2, for struts is D48t3.5.

All grids in the section are two-way spanning. The one-way spanning case is presented in Appendix D, giving an extensive understanding of the properties.

5.3.1 Design results

Chief design results are given in Table 5.1. In order to present the lightness cable-strut grids are compared with the DOS grid and the SOS grid. Two space grids are familiar forms and the DOS grid is generally the optimal. In addition, the results of the optimal tensegrity grid (contiguous strut grid of regular vertex-and-edge connection) in Table 3.3 are appended for comparison. Internal forces in the compressive and tensional layers that are characteristic of the load-transfer pattern of each grid are given in Figures 5.15–5.21 for reference. The distribution of forces in the SOS grid refers to Figure 3.17(b). In the figures, only one quarter of the grids are given due to the symmetry. Continuous lines refer to struts or bars, dashed lines to cables. The unit for forces is kN.

RP-a and RP-b grids

In the RP-a grid, the distribution of strut compressions is relatively even as a whole except for a few large values in boundary struts due to the forces transferred from boundary inclined cables (Figure 5.15(a)), which follows the property shown in the truss analysis. In addition, the distribution of tensions for both RP grids is also approximate to that in space trusses. The exception is that in the RP-b grid, the distribution of strut compressions is not even and small tensions occur in the central area (Figure 5.15(b)). The uniqueness seems mainly due to the torsional effect when struts are diagonal to the boundary (further study may be required to reveal why the effect is so 'severe'). However, such distribution can be adjusted by simple techniques such as camber design so that all struts are in compression with even values.

DP-a and DP-b grids

In contrast, DP grids present quite similar force distribution to space trusses, conforming well to the truss analysis results. In the DP-a grid, the force distribution (Figure 5.16(a)) is similar to that in the DOS grid (Figure 5.21) both under torsional effect. In the DP-b grid, the force distribution (Figure 5.16(b)) is similar to that in the SOS grid (Figure 3.17(b)). Also the DP-a grid is equivalent in lightweight properties to the RP-b grid (Table 5.1).

In the DP-b grid, stabilizing bars are subjected to small forces. They can be removed when semi-rigidity of joints is considered. Base cables simply

Table 5.1 Design results for various cable-strut and space grids

Grid types	*Parameters in design (m)*			*Max. internal forces (kN)*		*Weight (kg/m²)*		*Deflection under live load (span)*	*Numbers of joints/bars/cables*
	a	*h*	h_1 *or* h_u	*Compression*	*Tension*	*Strut*	*Cable*		
RP-a	3	2.5	0.75	182[a]	126	7.4	1.9	1/256	321/320/1160
RP-b	3.75	2.5	0.75	139[a]	162	5.9	1.6	1/311	272/320/834
CP-a	3.75	2.5	0.5, 1	119[a]	175	5.4	1.3	1/400	209/320/472
CP-b	3.75	3	0.5, 1	171	138	4.3	1.4	1/375	272/320/772
DP-a	3.75	2.5	1.5	99[a]	135	6.0	1.7	1/280	273/384/768
DP-b	3.75	2.5	1.5	178	175	5.3	1.6	1/284	336/412/992
ATP	3.75	2.5	0.5, 1	117[a]	121[a]	7.6	1.3	1/563	289/512/688
AP	3.75	2.5	0.5, 1	261[a]	154[a]	9.0	2.7	1/197	352/512/975
P-a	3.75	2.5	0.5, 1	184[a]	183	8.2	1.4	1/363	208/544/561
P-b	3.75	2.5	0.5, 1	183	76	6.7	1.6	1/448	352/512/1108
DOS	3.75	2.5	—	85[a]	163	12.3	—	—	208/652/0
SOS	3	2.5	—	140	132	14.0	—	—	221/800/0
TG	3	2.5	—	176[a]	286[a]	16.7	3.1	1/184	341/400/1020

Note

a Maximum value not in middle span.

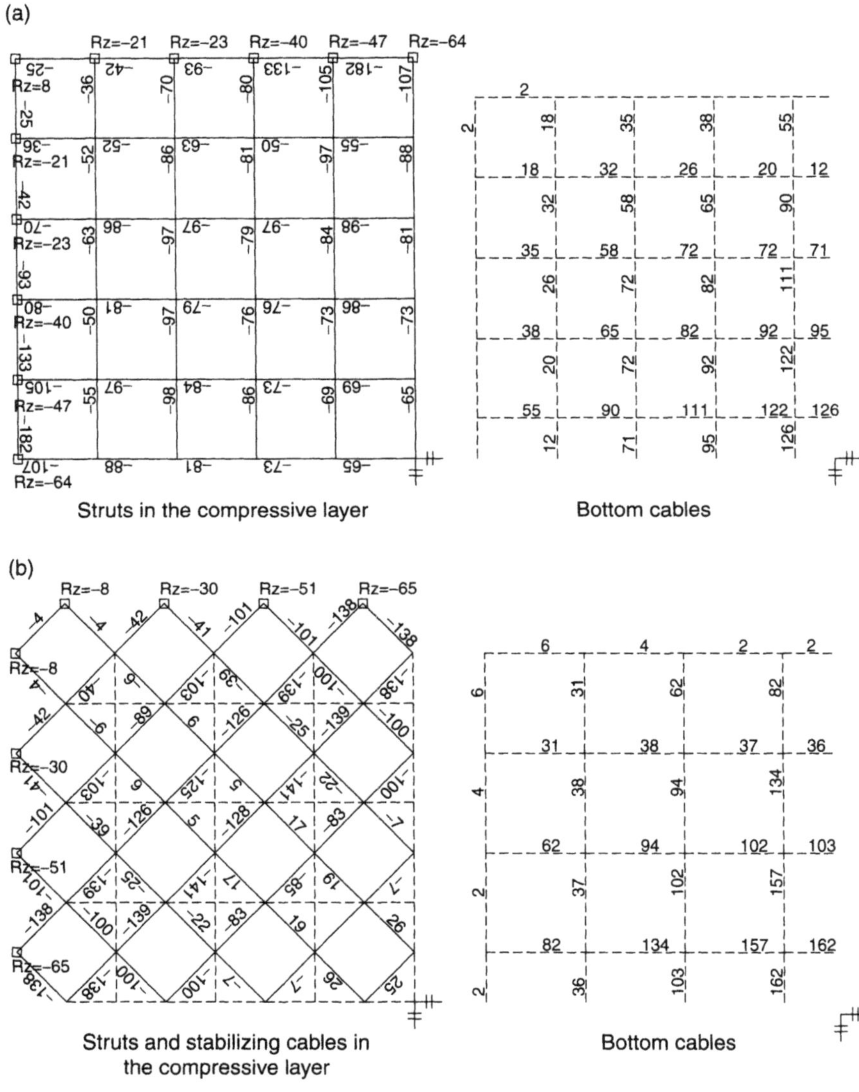

Figure 5.15 Internal forces of the RP grids: (a) RP-a grid; (b) RP-b grid.

Source: Wang and Li (2001); Courtesy: Multi-Science Publishing.

stabilize struts in the base under small forces. They may be further removed, if necessary, for example, in deployable model. In this case, the computation lengths of struts out of the vertical plane are doubled. Rectangular or elliptical tubes may be applied to balance the slenderness.

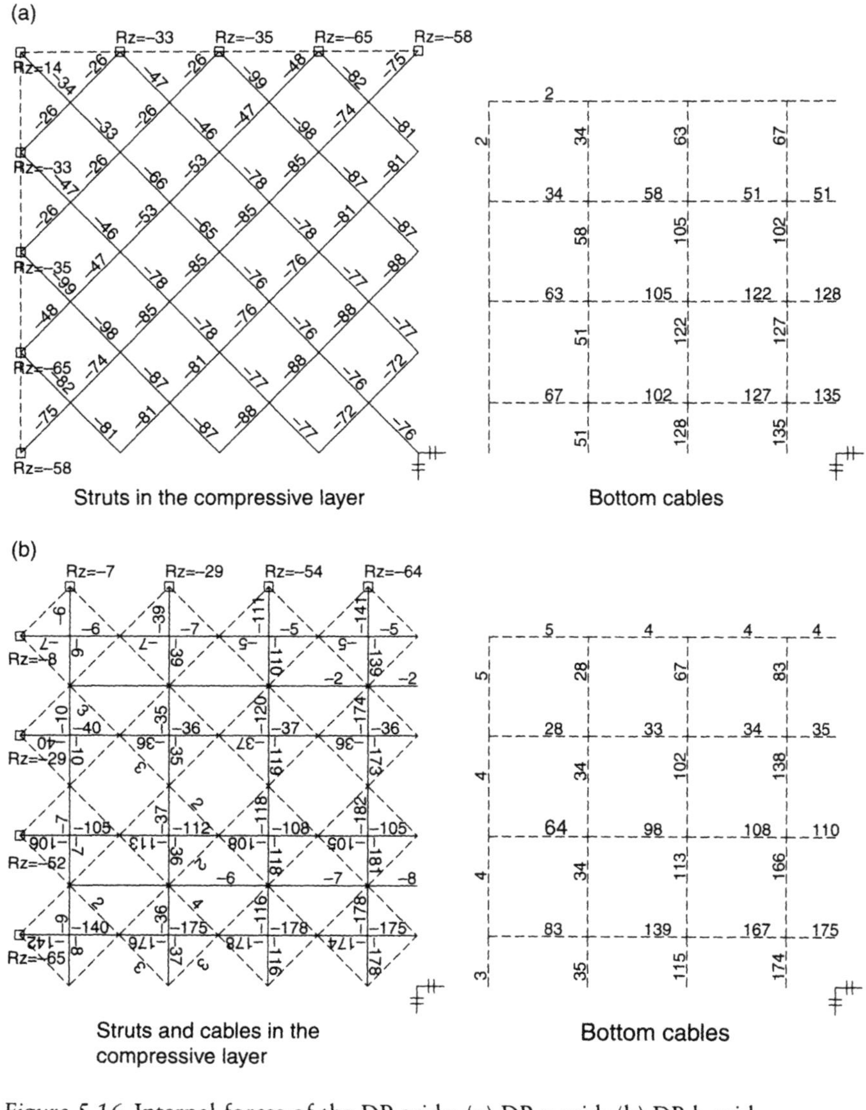

Figure 5.16 Internal forces of the DP grids: (a) DP-a grid; (b) DP-b grid.

CP-a and CP-b grids

As shown by the truss analysis, in both CP grids, upper cables in inner modules slacken or are subjected to very small tensions. In the CP-a grid, base cables in some inner modulus are removed. In the CP-b grid, base cables in inner modulus can also be removed when joints can take secondary moment and stabilizing cables can be further removed leading to double computation length of struts.

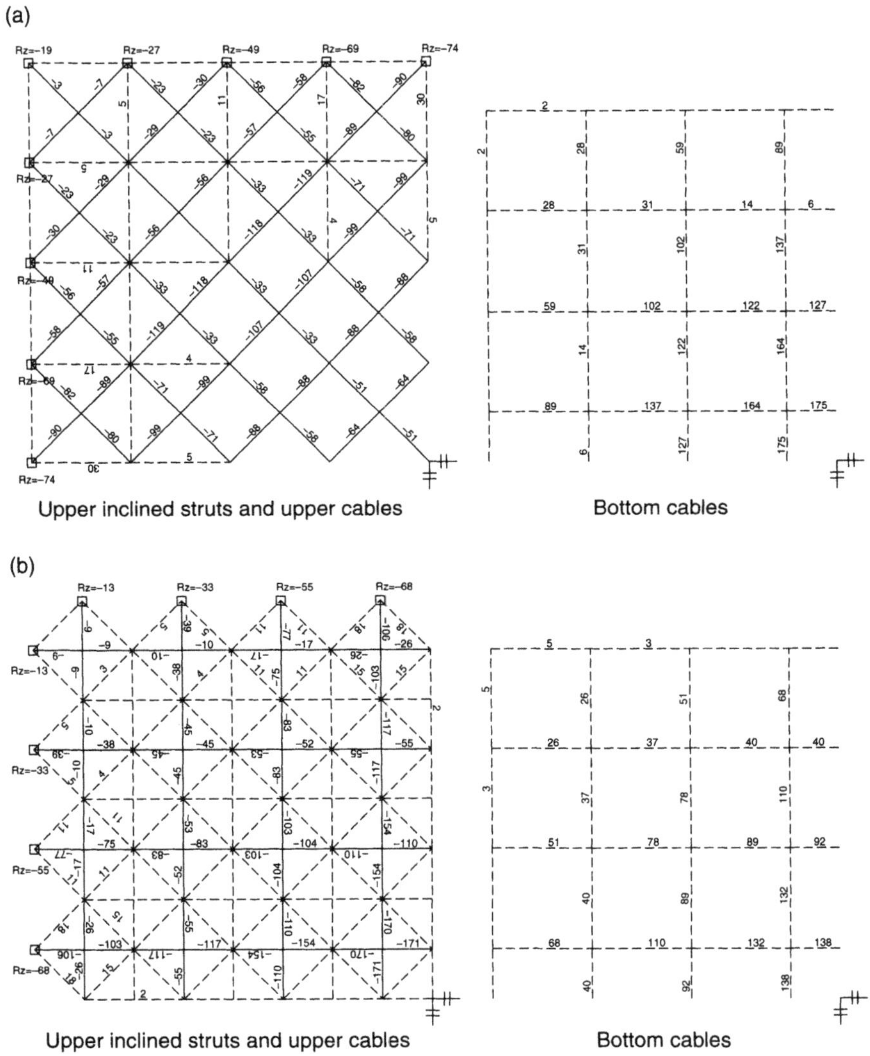

Figure 5.17 Internal forces of the CP grids: (a) CP-a grid; (b) CP-b grid.
Source: Wang and Li (2001); Courtesy: Multi-Science Publishing.

The force distribution is generally similar to that in DP grids. In general, the structural properties of the CP-a grid (Figure 5.17(a)) are compared to those of the DOS grid with torsional effect, and the CP-b grid (Figure 5.17(b)) to the SOS grid.

The strut forces in the CP-b grid (Figure 5.17(b)) are much larger than those in the CP-a grid (Figure 5.17(a)) although the depth is 1.2 times the

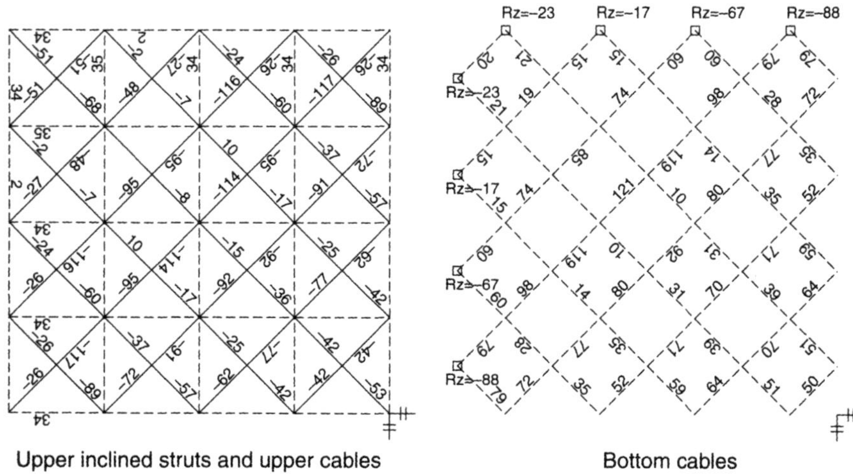

Figure 5.18 Internal forces of the ATP grid.

Source: Wang and Li (2001); Courtesy: Multi-Science Publishing.

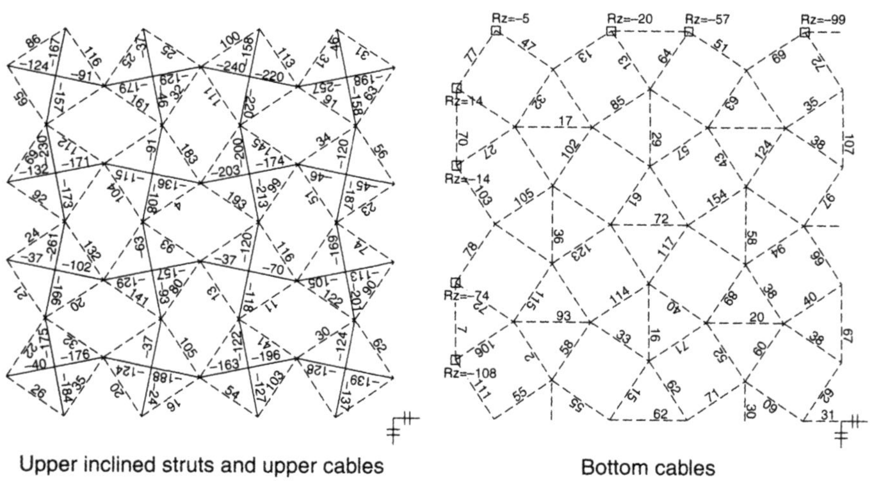

Figure 5.19 Internal forces of the AP grid.

latter. The main reason is that the modular length in the strut layer of the CP-a grid is $\sqrt{2}$ times the latter and that the latter benefits form torsional effect to achieve smaller forces.

ATP, AP and P grids

In the ATP grid, only upper cables in boundary modules are subjected to tensions (Figure 5.18). The overall property conforms with that obtaining

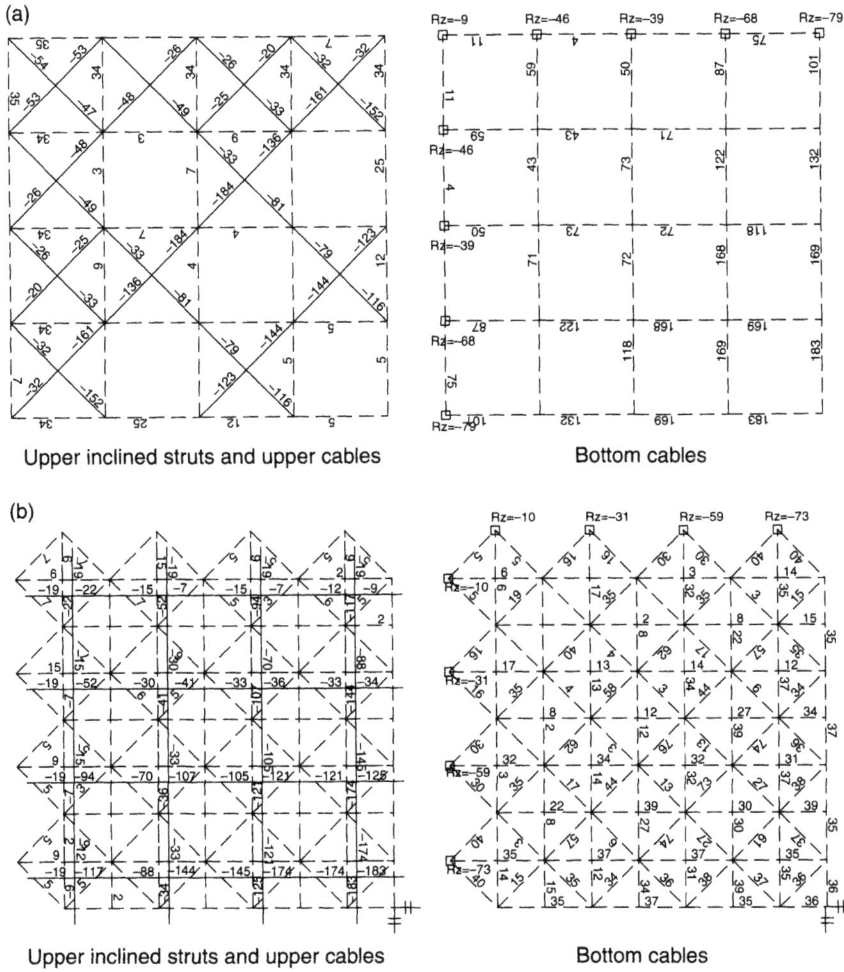

Figure 5.20 Internal forces of the P grids: (a) P-a grid (last quarter); (b) P-b grid.
Source: Wang and Li (2003a).

from the truss analysis and the overall distribution of strut forces follows the torsional effect.

In the AP grid, most upper cables are subjected to tensions producing larger compressions in struts as a whole (Figure 5.19). That is because each pair of struts is not co-planar in plan due to the relative rotation between two bases in each simplex (the factor cannot be expressed in the truss model), although the situation can be improved in the grid composed of the triangular simplex (Figure C.2 in Appendix C). In contrast, the force distribution in two P grids (Figure 5.20) is proved to be much more even and forces in upper cables are much smaller, conforming well with the property of the P truss.

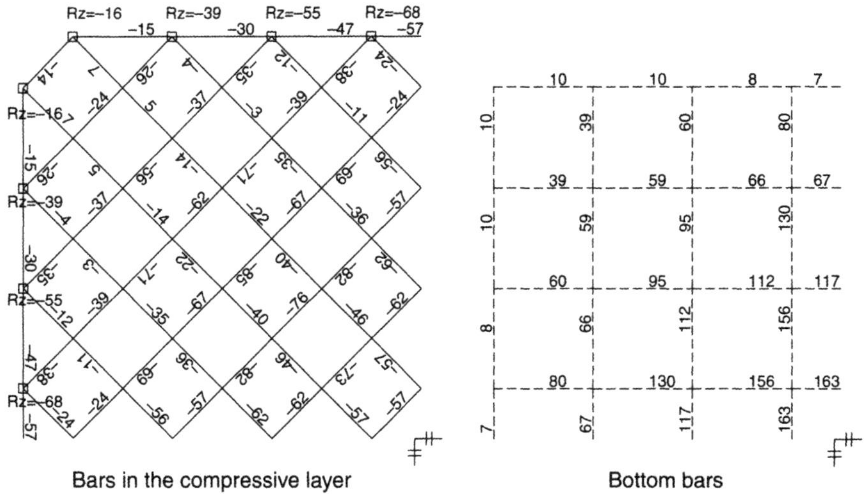

Figure 5.21 Internal forces of the DOS grid.
Source: Wang and Li (2001); Courtesy: Multi-Science Publishing.

P2 grids are similar in properties to those of P grids and so are AP2 (ATP2) grids to AP (ATP) grids, thus their results are not presented. They are in general less advantageous for conventional usage owing to increased cables and degrees of freedom, but their strut weight can be equivalently light.

Summary of internal forces

Based on the design results, truss analysis method can reveal the dominant load-transfer pattern of lightweight cable-strut grids. Generally, the influence of internal forces on structural efficiency is not so significant among lightweight grids, as the resistant level arms are already improved and optimal design can be applied.

It is interesting to note that average values of internal forces in the compressive and tensional layers in most grids are roughly proportional to load, modular length in the layer and inversely proportional to the resistant lever arms, but are relatively independent of grid types. Such property concerning some typical grids is summarized in Table 5.2 for reference.

5.3.2 Review on structural properties

In view of application, the CP grids are more suitable when the downward load case is dominant. The P-b grids and ATP grids are suitable in lighter roofs with restricted requirement on stiffness, and the P-a grids (with openings), RP grids and DP grids in more general cases. Structural properties are

Table 5.2 Average forces in the compressive/tensional layers for various cable-strut and space grids

Grid types	*Bars in the compressive layer*			*Tendons in the tensional layer*		
	Resistant lever arm (m)	*In-layer modular length (m)*	*Average compression (kN)*	*Resistant lever arm (m)*	*In-layer modular length (m)*	*Average tension (kN)*
Non-contiguous strut TG grid	< 1.25	3.4 – actual modular length	192	< 1.25	3.4 – actual modular length	76[a]
Contiguous strut TG grid	1.9	3	73	1.25	3	127
SOS grid	2.5	3	53	2.5	3	59
DOS grid	2.5	3.75/1.414	38	2.5	3.75	72
RP-a	≤ 2.5 – varies	3	81	2.5	3	52
RP-b	≤ 2.5 – varies	3.75/1.414	55	2.5	3.75	66
CP-a	1.9	3.75/1.414	60	2.5	3.75	64
CP-b	2.3	3.75	73	3	3.75	56
DP-a	≤ 2.5 – varies	3.75/1.414	69	2.5	3.75	64
DP-b	≤ 2.5 – varies	3.75	81	2.5	3.75	67
ATP	2.1	3.75/1.414	52	2	3.75/1.414	56

Note

a Due to crossing cables in the bottom layer, otherwise 2.5~3 times

evaluated as follows. Generally, the principles of inventing novel simplexes and lightweight grids turn out to be successful.

Weight reduction

The AP grid has the poorest properties compared with other cable-strut grids but still saves strut weight significantly due to bar length reduction compared with the optimal tensegrity grid. Other cable-strut grids are obviously lighter than space trusses. The ATP grid and P grids save more than half of self-weight compared with the optimal tensegrity grid and has good stiffness. The RP-b grid, two CP and two DP grids, owing to further reduction of bar density, can achieve more weight savings. The CP grids save strut weight mostly and the gross weight saving is nearly half compared with space grids (the CP-a grid with the SOS grid, and the CP-b grid with the DOS grid)!

The structural efficiency of cable-strut grids compared with space trusses can be illustrated more directly by appointing the same material grade for cables as for struts, here 200MPa so as to take out the material factor in structural efficiency. Hence, the weight of cables is increased to around 2.2 times. The final component weights, representing the 'actual' structural efficiency of cable-strut and space truss grids are compared in Figure 5.22. It shows the higher structural efficiency of cable-strut grids over space trusses due to the minimization of bars subject to buckling.

The experience of applying conventional cable networks impress people that weight reduction has little to do with economy. However, cable-strut grids may be exceptions. In addition to material saving, simplified joint design and convenience in construction are among other merits. The

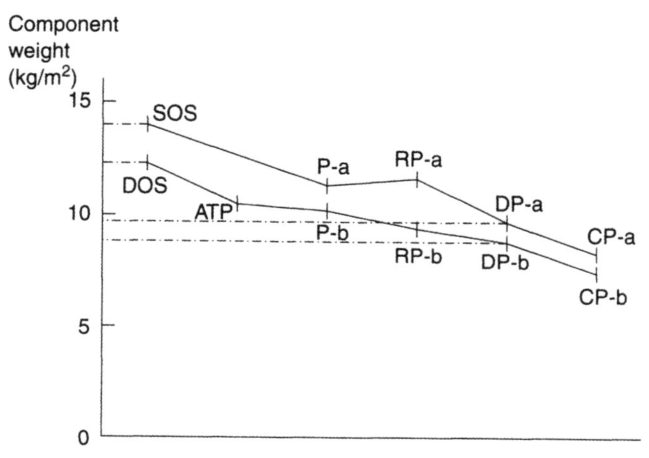

Figure 5.22 Comparison of structural efficiency based on the same material grade.

simplified joint design is presented in Section 6.2, and the convenience in construction is related to reduced weight, free-standing nature, and the fact that prestress is conceptually not necessary.

Stiffness

In addition to their lightness, the stiffness of cable-strut grids is also high. Ref. Table 5.1, the deflections under live load can meet the stiffness requirement for the conventional roofs (1/200–1/300 span) under non-prestress state or low prestress level. The ATP grid is especially stiff.

Grid depth for RP grids and DP grids

The present parameter design presents approximate resistant lever arms between cable-strut grids and space trusses for the convenience of comparing their properties. Therefore, RP and DP grids show higher overall grid depth ($h+h_u$). When it is an architectural requirement that both grids shall have reduced overall depth as space trusses, internal forces are increased. Self-weight is also increased but the proportion is less than 10% for the present cases. The 'worst' case, which seldom happens in practice, is that the design downward load and uplift load (after combined) are identical: h_u and h become half the depth in space trusses. The resistant lever arms are thus the smallest and the weight reduction becomes non-significant. For example, DP grids can only save 10% based on the same material grade for cables as struts.

6 Application studies of lightweight cable-strut grids

In this chapter, the large-span application, joint design and deployable functions of cable-strut grids are presented as further application studies. The technical information serves as the reference for design. However, the application of cable-strut concept should not be limited by these aspects and are yet to be exploited. For example, the hybrid forms of cable-strut grids with other structural types may be used in the design to combine their merits and it will challenge the wisdom of designers and engineers.

In large span applications, large internal forces, complex construction process, joint design, weight reduction and structural analysis, etc. are among issues that need advanced techniques. By reference to the analysis in Chapter 5, tensegrity grids and the AP grid composed of square simplexes are not preferred due to their large self-weight load and large internal forces, which produce difficulties in joint design. In comparison, lightweight grids of smaller internal forces are suitable in super spans when self-weight load becomes significant. Among them, the DP-a grid is studied representatively. The spirit can be extended to other grid types. Besides, domical forms are not a focus in the present study, but cable-strut grids shall conceptually present good performance.

Joint design is important for any structure as it may affect the structural performance, the aesthetics, the fabrication price and the assembling process in construction, etc. On the premise of efficiently transferring internal forces, joints are required to be light, easy to fabricate, fast assembling, good looking, etc. It is preferred that the joint design can be as simple as possible so as to meet these requirements. In cable-strut grids, cable connection is simple and struts are subjected to compression. The principles of the joint design are recommended in this chapter.

Deployable structures hold full volume as supporting structures when deployed and have smallest volume for transportation when collapsed into a compact bundle. The main benefits of these structures are that they are reusable, can be kept with very small volume, and can be assembled and released very quickly with least manual or mechanical forces. Most 'conventional' deployable structures are composed of bars. In comparison, cable-strut systems can achieve less stowed volume (higher packaging efficiency) and reduced weight (higher structural efficiency), and decrease the number of joints that have complicated mechanisms. Discussion of deployable functions includes telescopic strut method, energy-loaded strut method and releasing cable method, etc. among possible means. Finally, the retractable functions are mentioned.

6.1 Large span design

In this section, large span design of DP grids is presented for reference. Two examples are given as follows.

6.1.1 Example one – 100m span, circular layout

The grid is of circular plan (100m span) and simply-supported at the boundary nodes in the middle layer. That is to say, the supporting structures for the grid are not supposed to sustain lateral forces, unlike conventional cable networks. For the DP-a grid itself, most roof materials including membrane or metal panels are suitable when secondary members are introduced. The light membrane is used in the design in order to consider the effect of suction wind load. Note that all cable-strut grids have upper cables, with which membrane can be stiffened so as to avoid large stress and deformation under uplift load.

Load cases

Load cases are given as follows.

Roof dead load: $P_d = 30\text{kgf/m}^2$
Live load: $P_l = 70\text{kgf/m}^2$
Suction load: $P_s = 100\text{kgf/m}^2$

Here, the roof dead load includes the weight of the membrane, the secondary cables for attaching the membrane and prestress load in membrane, excluding the self-weight load of the grid (P_{sw}). Other load types, such as seismic load, are not considered as they do not take effect often. But it is of interest to include it in further study.

Based on the given loads, two combined load cases are considered. One is the full load case (downward), the design value based on Chinese code

$$P_f = 1.2\ P_d + 1.2\ P_{sw} + 1.4\ P_l \tag{6.1}$$

The other is the uplift load case, the design value

$$P_u = 1.4\ P_s - P_d - P_{sw} \tag{6.2}$$

Assuming that the self-weight $P_{sw} = 30\text{kgf/m}^2$, then

$$P_f = 1.2 \times 30 + 1.2 \times 30 + 1.4 \times 70 = 170\text{kgf/m}^2 \tag{6.3}$$

$$P_u = 1.4 \times 100 - 30 - 30 = 80\text{kgf/m}^2 \tag{6.4}$$

These loads are distributed by area to all upper joints. As such grid is not sensitive to unsymmetrical loads, half-span load cases are not considered.

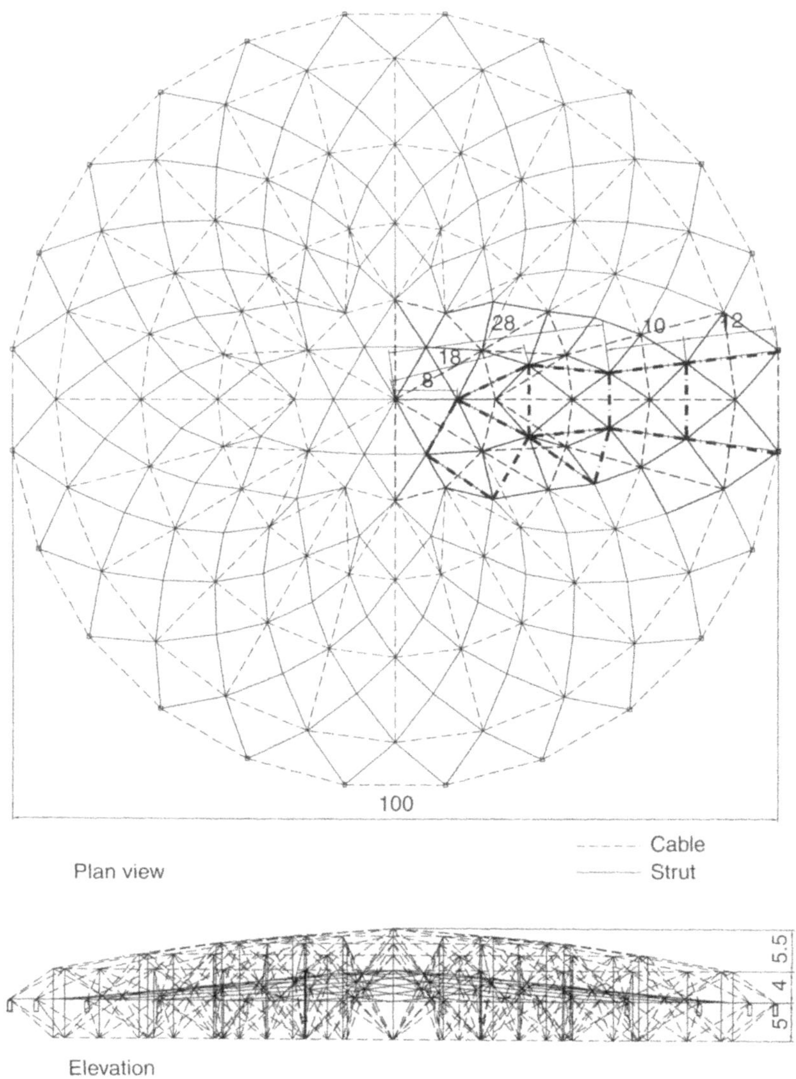

Figure 6.1 General layout of the DP-a grid in 100m span.

Grid description

The primary concern of grid design is to avoid densified components and make even the distribution of internal forces. The resulting self-weight may not be the optimal. By referring to the general layout in Figure 6.1 and the layout of the strut layer in Figure 6.2, the modules vary from large trapezoids in the outer ring to smaller trapezoids in the second outer ring, then to two

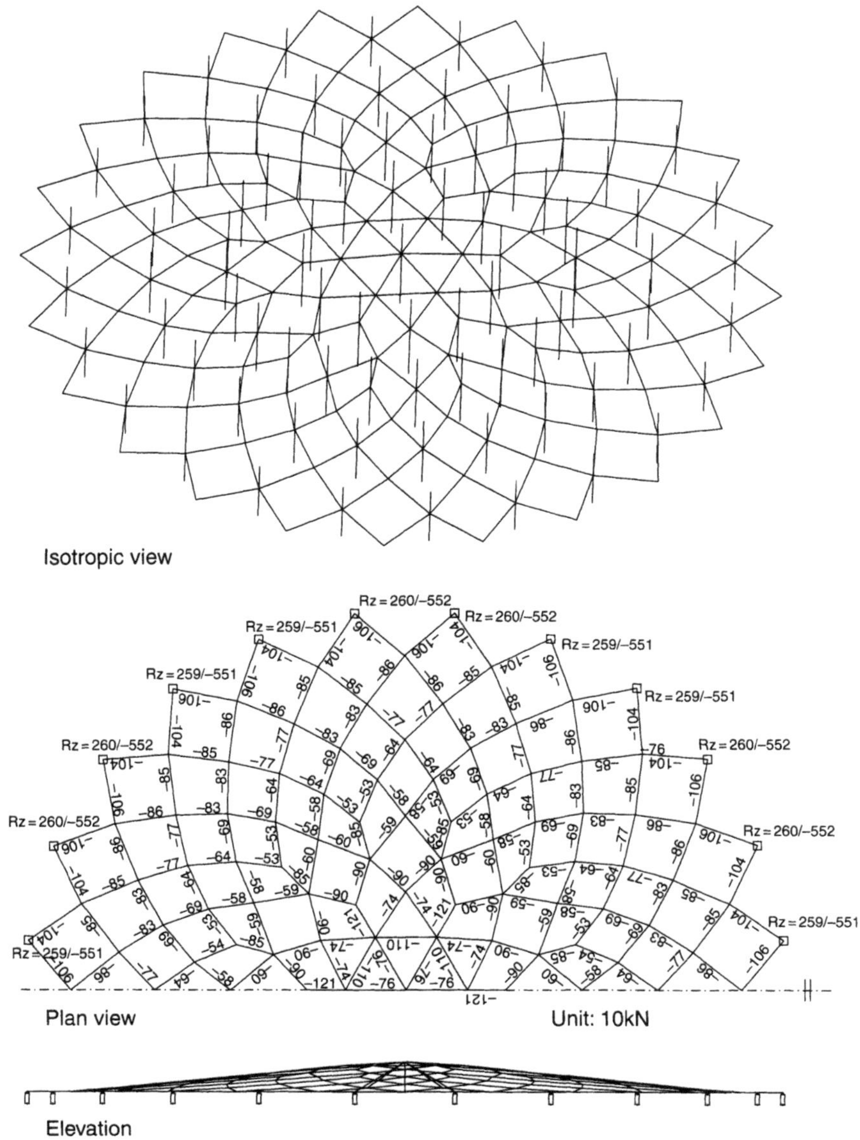

Figure 6.2 Layout and internal forces of the strut layer of the DP-a grid in 100m span.

rings of mixed trapezoids and triangles, finally, to a hexagon at the centre. The number of modules is nine across the span. The modular lengths vary in radial direction from the edge to the centre: 12m, 10m, 10m, 10m and 16m. The rise for the upper layer is 5.5m, presenting the span-to-rise ratio of 16:1.

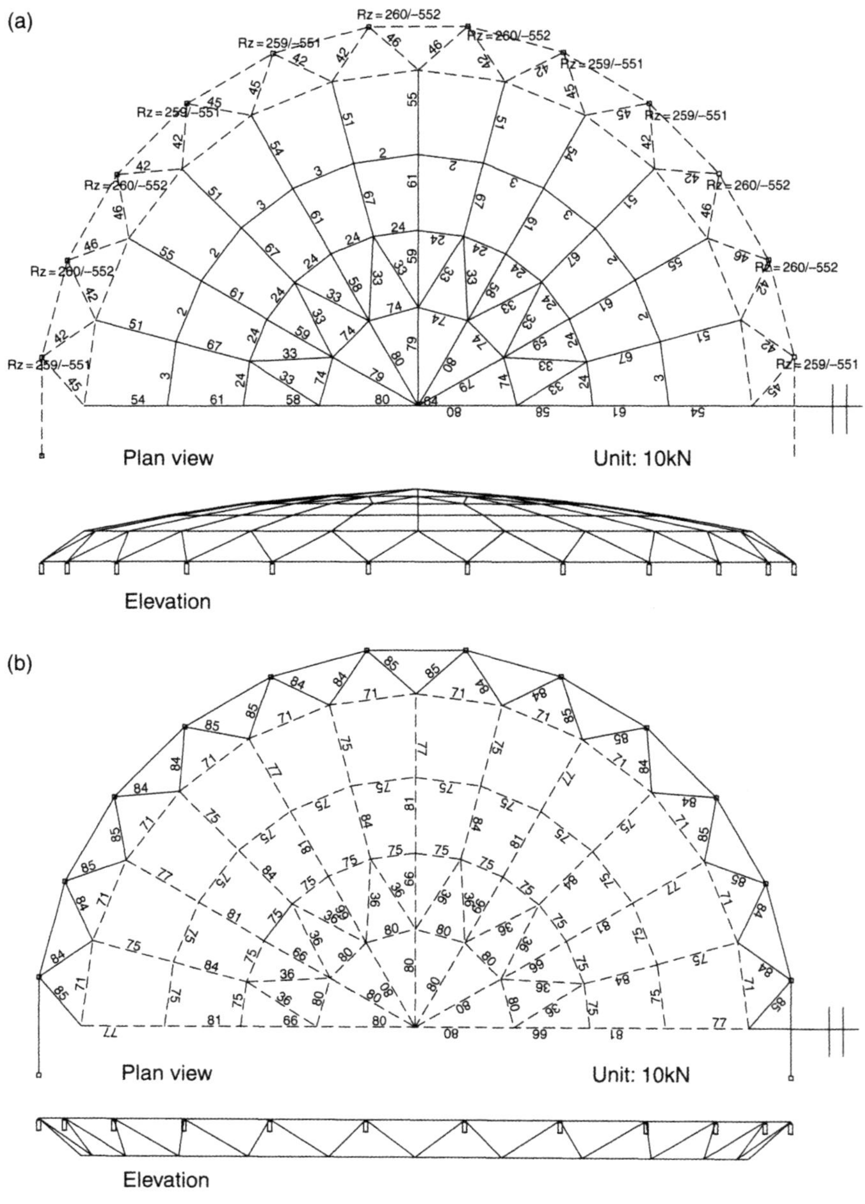

Figure 6.3 Layouts and internal forces of cables of the DP-a grid in 100m span: (a) upper layer; (b) bottom layer.

The rise for the compressive layer is 4m and the sag for the bottom layer is designed to be 5m. When necessary, the overall grid depth can be reduced.

The grid contains 500 struts and 1,032 cables. The layout of upper tensional layer with boundary upper inclined cables is presented in Figure 6.3(a), and the layout of bottom tensional layer with boundary lower inclined cables in Figure 6.3(b). Following the layout of the upper tensional layer, the membrane is cut into seven sizes only.

Design results

In the design, material strength for struts is 200Mpa, and for cables 500Mpa. Tubes for designing struts are D194t5, D219t6, D245t7 and D299t8. Cable cross-section in design are 0.0003m^2, 0.0006m^2, 0.0012m^2 and 0.0018m^2. No prestress is introduced. The resulting overall deflection under uniform suction load alone (without being offset by dead load) is 1/100 span. Deflection under uniform live load only is 1/290 span.

Under the two combined load cases, the supports sustain compression and tension respectively. The bottom layer and the upper layer act as the tensional layer alternatively whereas the strut layer always functions as the compressive layer. The distribution of compression is given in Figure 6.2 for reference. The maximum value 1210kN, also the largest of the grid, lies in the triangular modules at the second ring from the centre.

The distribution of tensions in the upper and bottom layer respectively is illustrated in Figure 6.3(a) and (b). The maximum tension is 850kN, lying in lower inclined cables connected to the boundary. The forces are distributed quite evenly as a whole, so joint design can be easier by clamping continuous cables to nodes. It is interesting to note that in central area, the tensions for both layers are almost identical. This is due to the balanced design of their resistant lever arms according to the proportion between the design upward and downward load values in Eqns (6.3) and (6.4).

Finally, the unit weight for struts (based on centre-to-centre length) is 18kg/m^2, for cables 8kg/m^2, altogether, 26kg/m^2. It is much lighter than the space truss of the similar layout, about 45kg/m^2.

6.1.2 *Example two – 200m span, square layout*

The grid is of square plan (200m × 200m) and simply-supported. The roof material covering the outer layer, for such a large span, is prescribed membrane, and sufficient rise is required. In such a large span, internal forces are always large hence base-reinforcement method is introduced first to reduce internal forces.

Base-reinforcement method

The base-reinforcement method is originally applied into space trusses to reduce internal forces when the increase of grid depth is not preferred

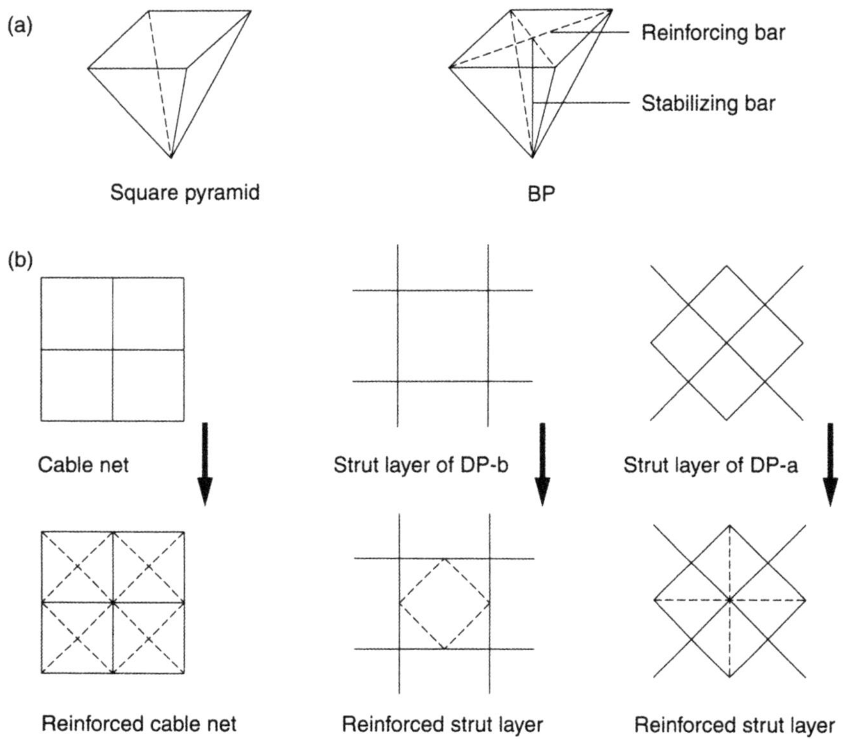

Figure 6.4 (a) Base-reinforcement principle (source: Wang 1998c); and (b) derived methods.

(Wang 1998c). Meanwhile, such method has additional merits in the improvement in stiffness and stability. Following the principle, four co-planar bars with a common node are introduced into the base of each square pyramid module (Figure 6.4(a)). In this condition, an additional bar is needed to stabilize them. The method reduces internal forces in space trusses significantly (up to two-thirds, Wang 1998c) without densifying modules and increasing many web elements.

The case is the same for the 200m grid here. The internal forces are quite high but it is not a good practice to increase grid depth arbitrarily. The applications of base-reinforcement principle to DP-a and DP-b grids are shown in Figure 6.4(b). But in the present design of the DP-a grid, only cables in the bottom layer are reinforced. As cables are always in tension, no stabilizing component is needed.

Grid description

In designing the DP-a grid, the number of modules in each way is odd, nine, in consideration of water drainage at the centre module. The resulting modular length is 22.222m. The rise for the upper layer is 11m, giving the span-to-rise ratio of 16 : 1, which is conventional for membrane roofs. The rise for the compressive layer and the sag for the bottom tensional layer are both 10m. The general layout and the isotropic view of the layout are shown in Figures 6.5 and 6.6, respectively.

Based on the design, the grid contains 486 struts. The lengths of struts in the compressive layer are around 15.6m. The lengths of the upper vertical struts are equal, 10m, except for the central one, 11m, those of the lower

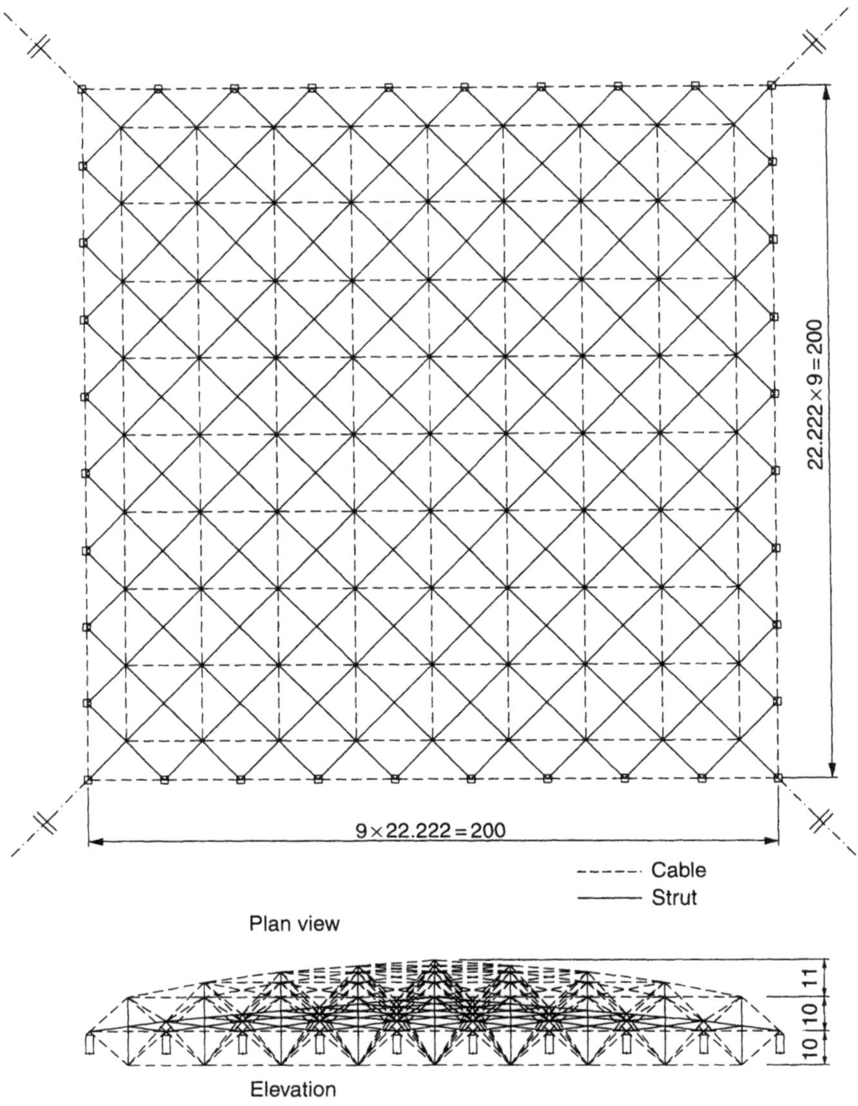

Figure 6.5 General layout of the DP-a grid in 200m span.

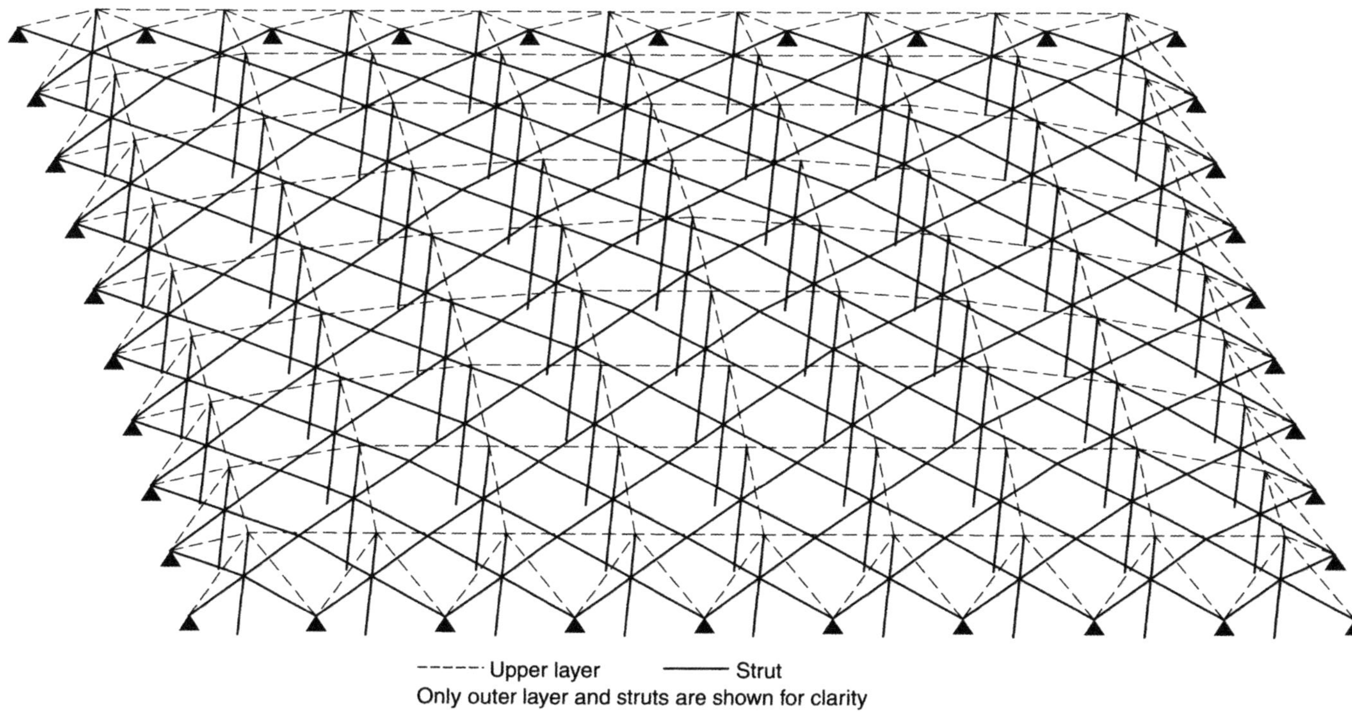

Figure 6.6 Isotropic view of the DP-a grid in 200m span.

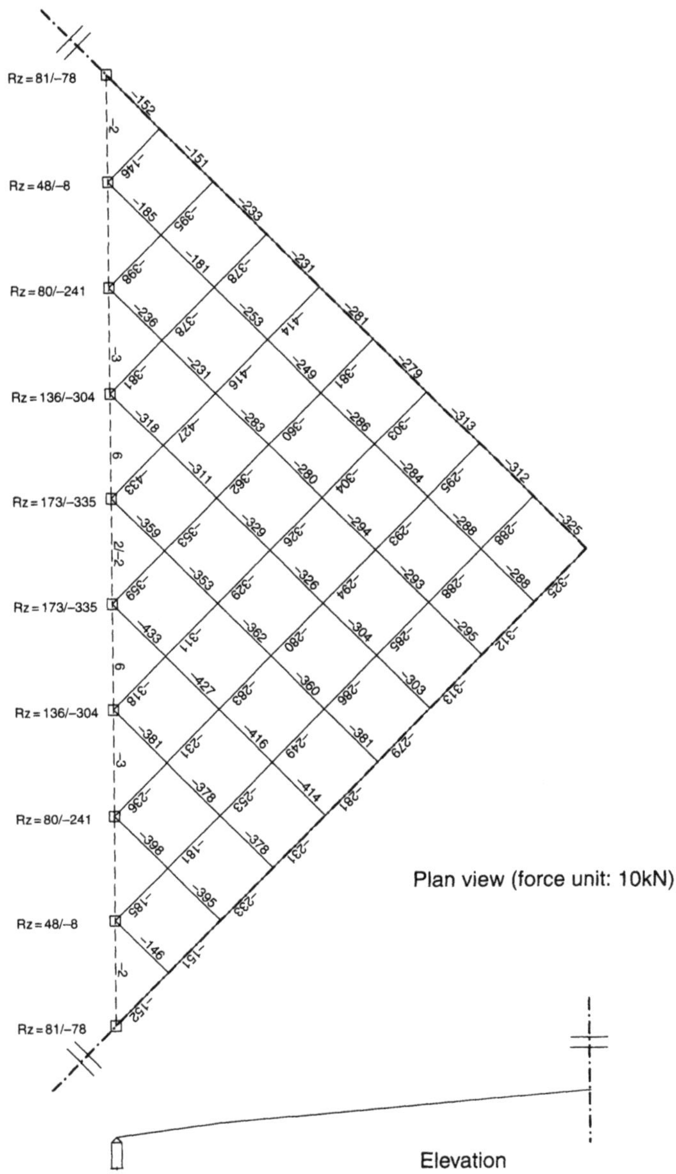

Figure 6.7 Shape and internal forces of the strut layer of the DP-a grid.

vertical struts ranging from 10m to 20m. A quarter of the compressive layer is illustrated in Figure 6.7.

The grid contains 1,100 cables. Among them, crossing cables are introduced in order to reduce cable tensions in the bottom layer. These cables can rigorously share tensional forces and make even the force distribution

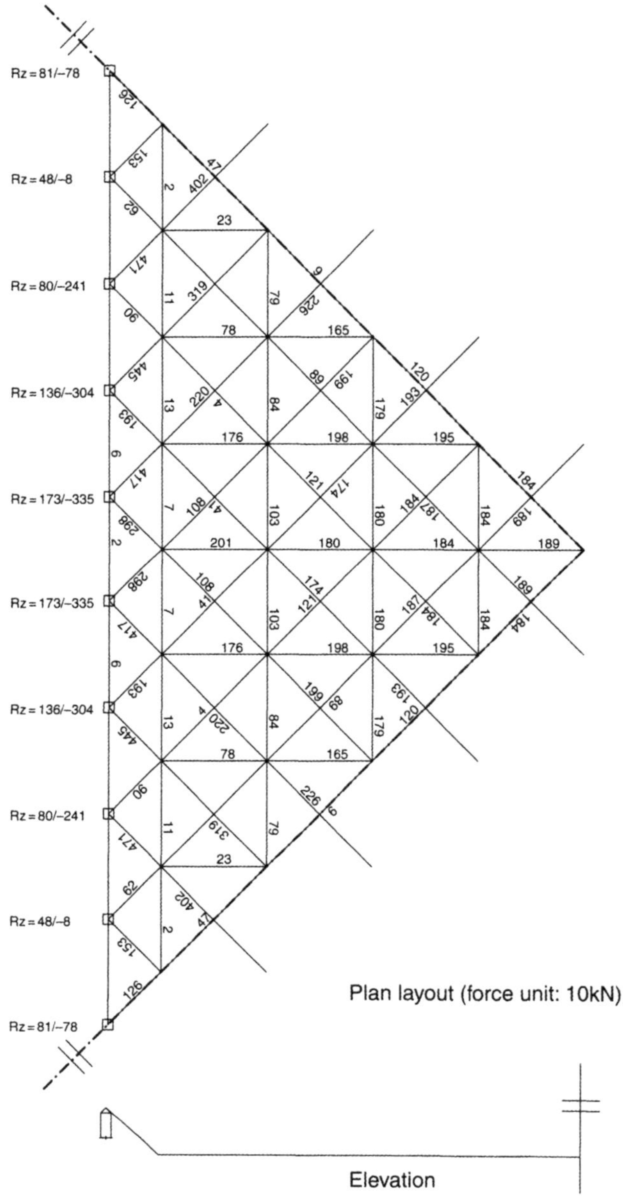

Figure 6.8 Shape and internal forces of the bottom layer of the DP-a grid.

by reducing the maximum cable tension in the layer by two-thirds. A quarter of the bottom tensional layer is given in Figure 6.8.

A quarter of the upper tensional layer is shown in Figure 6.9. Based on the design, membrane as a whole is cut into only eight different sizes. Upper cables

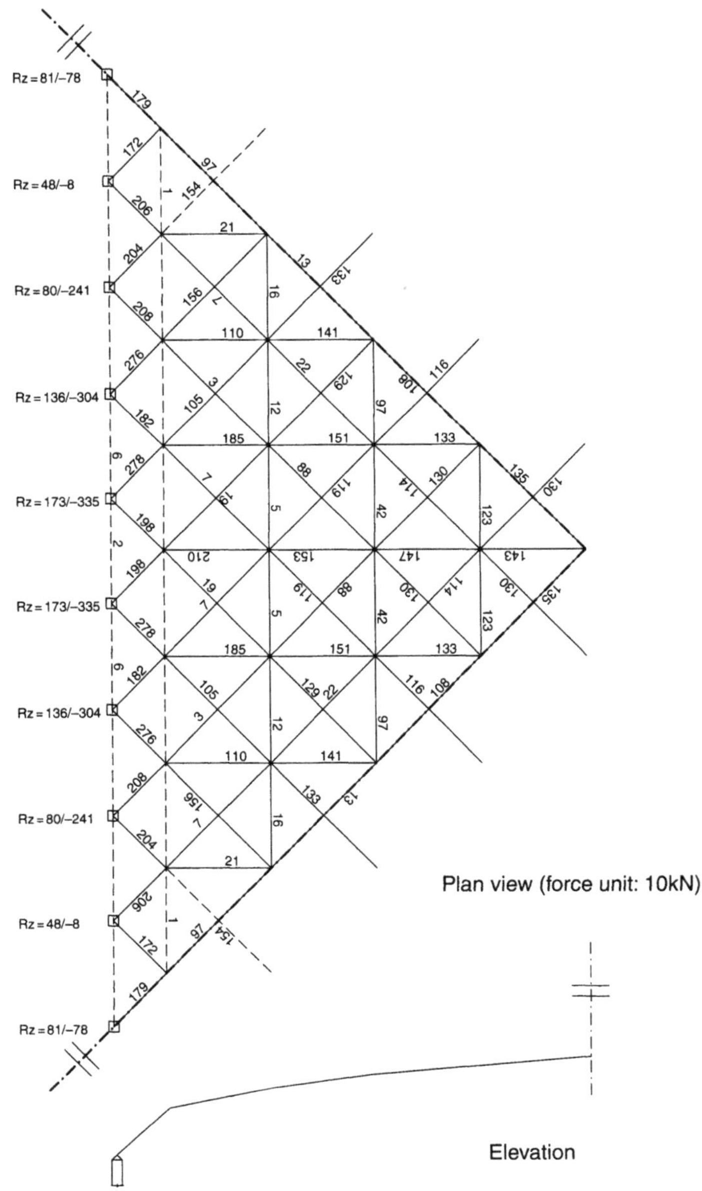

Figure 6.9 Shape and internal forces of the outer layer of the DP-a grid.

actually function together with membrane roofs but the effect is not considered here. Note that introducing crossing cables can reduce the size of cable nets for attaching membrane and also the depth of the upper part since cable forces are much reduced, but it is not considered for simplicity.

Design results

As span is large, high-strength steel is introduced. Here, the design material strength for struts is 300MPa, for cables is 1,000MPa. Tubes for designing struts are D325t8, D377t9, D450t9, D530t9, D600t10 and D600t12. Cables cross-sections in design are 0.0005m^2, 0.001m^2, 0.002m^2, 0.003m^2, 0.004m^2 and 0.005m^2.

For such a large span, it is preferred to introduce a low level of prestress to improve the stiffness and stability of the whole grid. It is assumed that 500kN compression is introduced to all lower vertical struts. Such level of prestress has little influence on self-weight and internal forces as a whole. The effect of membrane and its prestress to upper cables not considered, the resulting maximum deflection under uniform suction load alone (without being offset by dead load) is around 1/65 span. In comparison, the maximum downward displacement under uniform live load only is 1/200 span.

The strut layer always functions as the compressive layer. The distribution of compression is given in Figure 6.7. It shows torsional effect, and the maximum value is 4,330kN, which is the largest of the grid. The largest compression in the vertical struts is about 4,000kN, lying in the lower vertical struts in some boundary modules.

Under the two combined load cases, the bottom and upper layers act as the tensional layer alternatively. The distribution of final tensions in the upper and bottom layer is illustrated in Figures 6.8 and 6.9, respectively. The maximum tension is 4,710kN, lying in lower inclined cables connected to the edge. But in the majority of cables, the tensions are relatively small in view of such large span. For a few cables of large tension, both ends are connected to only one strut hence the joint design ought not to be difficult. In contrast, large tension is in general a problem in large span design.

The lightweight of the present design is prominent at the cost of high grid depth and high-strength material. The unit weight for struts (based on centre-to-centre length) is 20kg/m^2, for cables 9kg/m^2. Including the joint weight, the total weight is only about 30kg/m^2!

6.1.3 *Summary and discussion*

Design parameters and self-weight

The DP grids achieve optimal lightness at relatively high grid depth. In the present examples, the overall depth for Example One is 14.5m for 100m span, for Example Two 31m for 200m span. The depth can be reduced with the increase of weight for the large design modules. For example, when we proportion the depth from 14.5m to 10m in Example One, the resulting component weight increases from 26kg/m^2 to 32kg/m^2.

Besides, designing large modules with large tube diameter and grid depth proportional to span leads to the effect that self-weight does not increase proportionally with span. In Example One, if we double all dimensions (the resulting module lengths and grid depth are then approximate to those in Example Two) and utilize the same cross-sections and high strength material as in Example Two, the resulting component weight increases from 26kg/m^2 in 100m span to 32kg/m^2 in 200m span, 10% heavier than the square layout in Example Two. When there is the restriction on overall depth, smaller modules can be used to control internal forces.

Comparison with cable domes

Present studies show that it is not difficult for cable-strut grids to achieve below 40kg/m^2 in 200m span. Cable domes are lighter but are actually not highly structurally efficient whose weight reduction is due to high strength of cables. In comparison, cable-strut grids save a boundary ring beam and avoid complicated construction process.

Cable-strut grids extend the merits of cable structures by the incorporation of simple techniques in space trusses. The dominant advantages of cable-strut grids over cable domes lie in two aspects. One is that joints in cable-strut grids can be standardized and cheap (as discussed in the following section), whereas joints in cable domes cannot be standardized and are quite expensive. The other is that the former are much simpler in construction in that they are free-standing and do not require high prestress. In contrast, the latter are not free-standing and rely on high prestress to prevent the slackening of the majority of cables so as to maintain stability and sufficient stiffness. So in large spans, cable-strut grids can be much more economical than conventional cable networks.

Discussion on domical forms

When the DP grid in Example One is applied into higher-rise domical form, grid depth is much reduced. As the property is simple, details are omitted and only the conclusions are presented. Here, it is sufficiently stiff to apply the present single-layer strut form in 100m span domes. Application of double-layer and triple-layer struts refers to the discussion in Section 4.6.

Compared with flat forms, forces in domical forms are transferred to the edge, namely, thrust forces, which are positive under downward load and negative under uplift load. The values vary with rise but are normally quite big. When the thrusts are sustained by the boundary ring beam or supports, the in-span weight is reduced quite significantly. Adjusting the profile (rise) can reduce the thrust but does not have much influence on total weight. The behaviour is generally similar to that of a latticed bar shell and is general for all cable-strut domes.

Generally, bars in the middle layer (except those at the edge) of the DP grid are compressive under downward load but are tensional under suction load. Cables mainly act as stabilizing components and the weight contribution is quite low. Extended to other cable-strut domes, RP grids are similar in properties. The left three types (AP, ATP and P domes) are stiffer, containing two layers of inclined struts.

6.2 Joint design in lightweight cable-strut grids

6.2.1 Components and joint types

In actual construction, there are many options available for designing components. For struts, circular/elliptical or square/rectangular tubes, timber bars, and even air-inflated plastic pipes, etc. are among the possible options. For cables, tension rods, steel strands, steel ropes, flat bars and even tubes are feasible. A steel tube, as an idea, has two flattened ends, each bolted to a plate with a slot hole (Figure 6.10(a)). The bolt can slide in the slot hole to release compression. The idea for the flat bar is similar (Figure 6.10(b)). *Based on the principle, the purlin member, when necessary in metal roofing, can be applied as the top tensional member simultaneously.*

There are a wide variety of choices in designing joints for cable-strut grids. For example, joint design in tension structures shall be indicative, such as the tubular joint in glass cladding (Figure 6.11). But as cable-strut grids are basically a family of space frames, their joint design may conceptually refer to another family of space frames, space trusses, as studied here.

In space trusses, there are mainly four types of prevailing joint systems: solid bolted spherical joints, hollow bolted spherical joints, welded hollow spherical joints and plate-type joints. In applying four types of joint systems into cable-strut grids, welded hollow spherical joints require site welding. Other three types are discussed in the following sections.

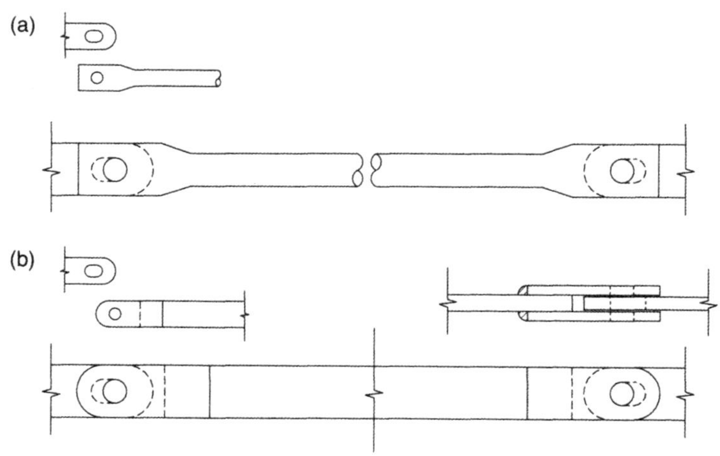

Figure 6.10 Tensional elements: (a) tubular element; (b) flat bar.

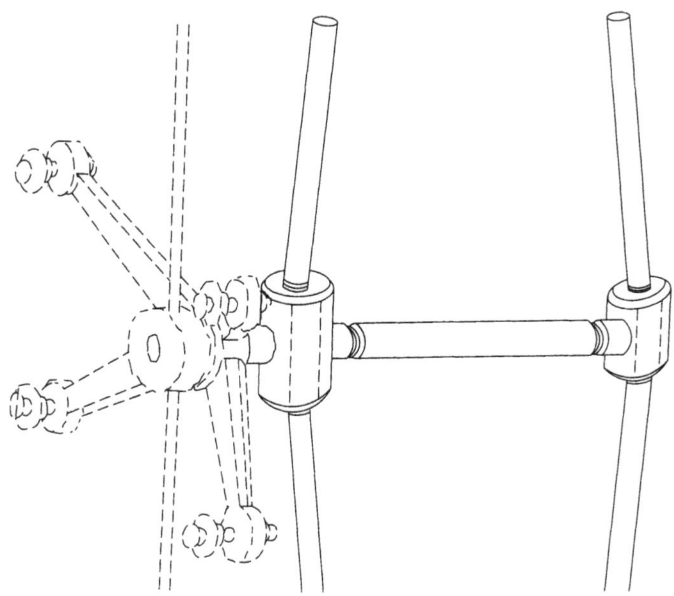

Figure 6.11 Tubular joint in glass cladding.

6.2.2 *Solid bolted spherical joint system*

Solid bolted spherical joints are used frequently in space trusses. In connecting a tube to the bolted ball, it requires a bolt, a nut, a pin and a cone welded to the tube (Figure 6.12(a)). The connection method can also be used for the struts in cable-strut grids. As struts are mainly subjected to compression, the bolt can be smaller as little force is transferred to it.

In comparison, the connection of a tendon to the ball is much simpler and requires a small contact area (Figure 6.12(a)). Only an additional turnbuckle is required to tighten it. The simplicity is an important benefit of applying tendons over tensional tubes in space trusses. The appearance of applying tendons and solid bolted joint in a DP grid is presented in Figure 6.12(b).

Solid bolted spherical joints are simple and can express the simplicity of connecting tendons. However, the joints have the drawbacks in that the diameter of the joints has to be increased when the included angle between any two components is small. At this aspect, hollow spherical joints with larger outer diameter are more advantageous.

6.2.3 *Hollow spherical joint system*

A typical hollow bolted ball can be designed as a half ball (Figure 6.13(a)), as in the upper and bottom layers of various cable-strut grids. In connecting the tendon to the hollow ball, the nut is applied inside the ball to fasten

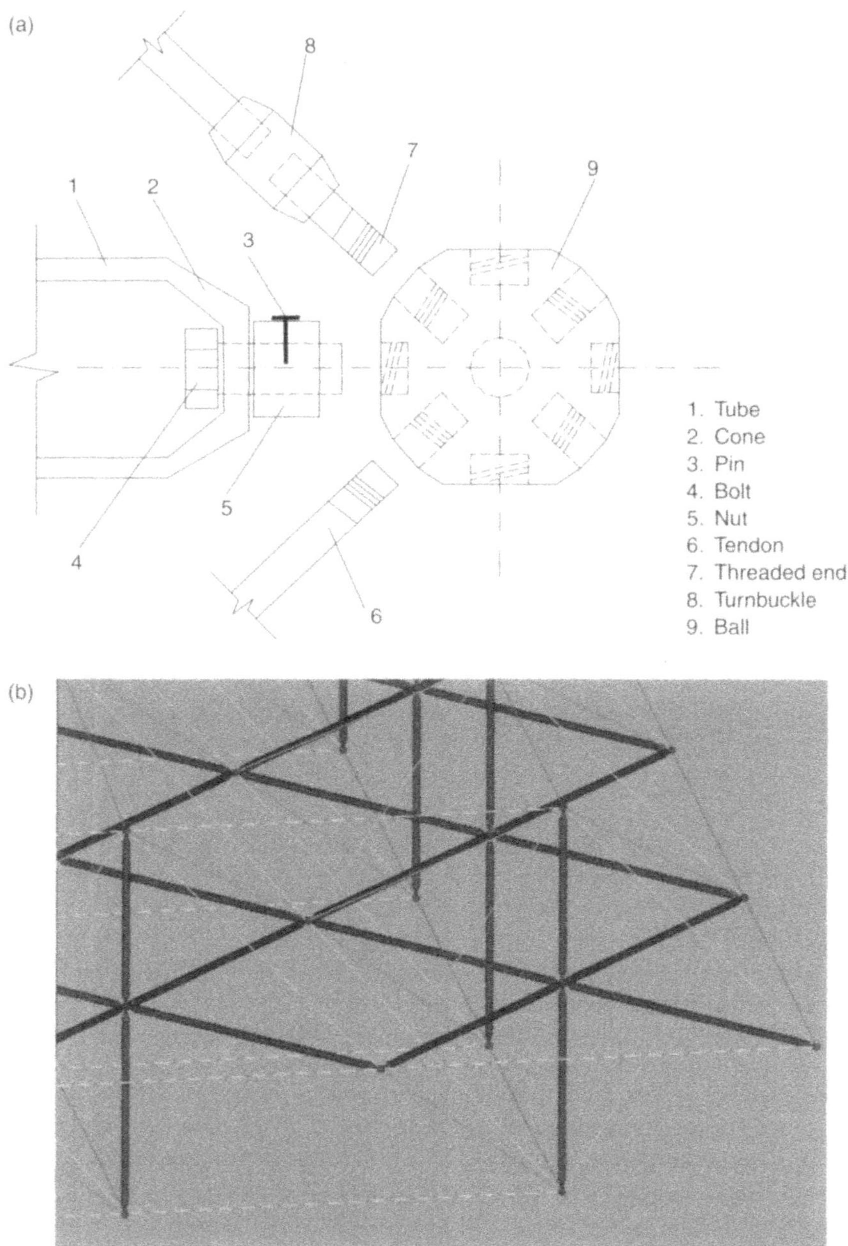

Figure 6.12 Solid bolted spherical joint system: (a) joint design; (b) the resulting DP-a grid.

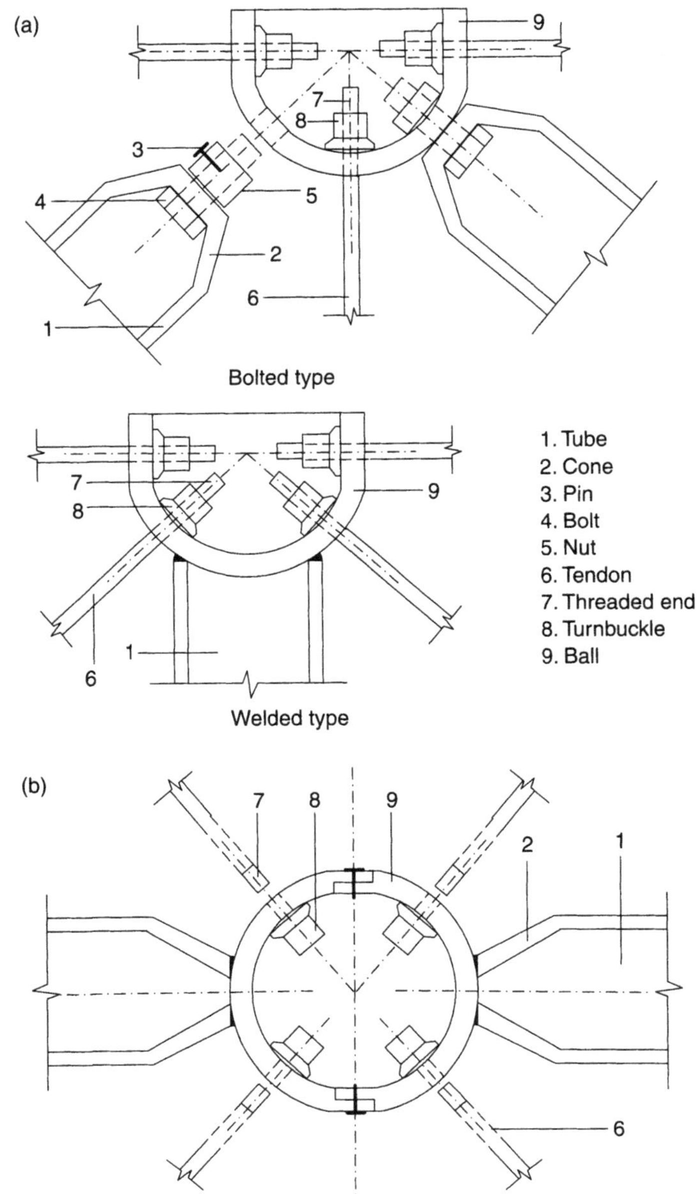

Figure 6.13 Hollow spherical joints: (a) hollow spherical joint system; (b) hollow spherical joint in the middle layer with cables.

the threaded end, thus a turnbuckle is not necessary. In connecting the tube to the hollow ball, the connecting assembly can be the same as the solid ball, but the small nut can also be fastened from inside the ball (Figure 6.13(a)).

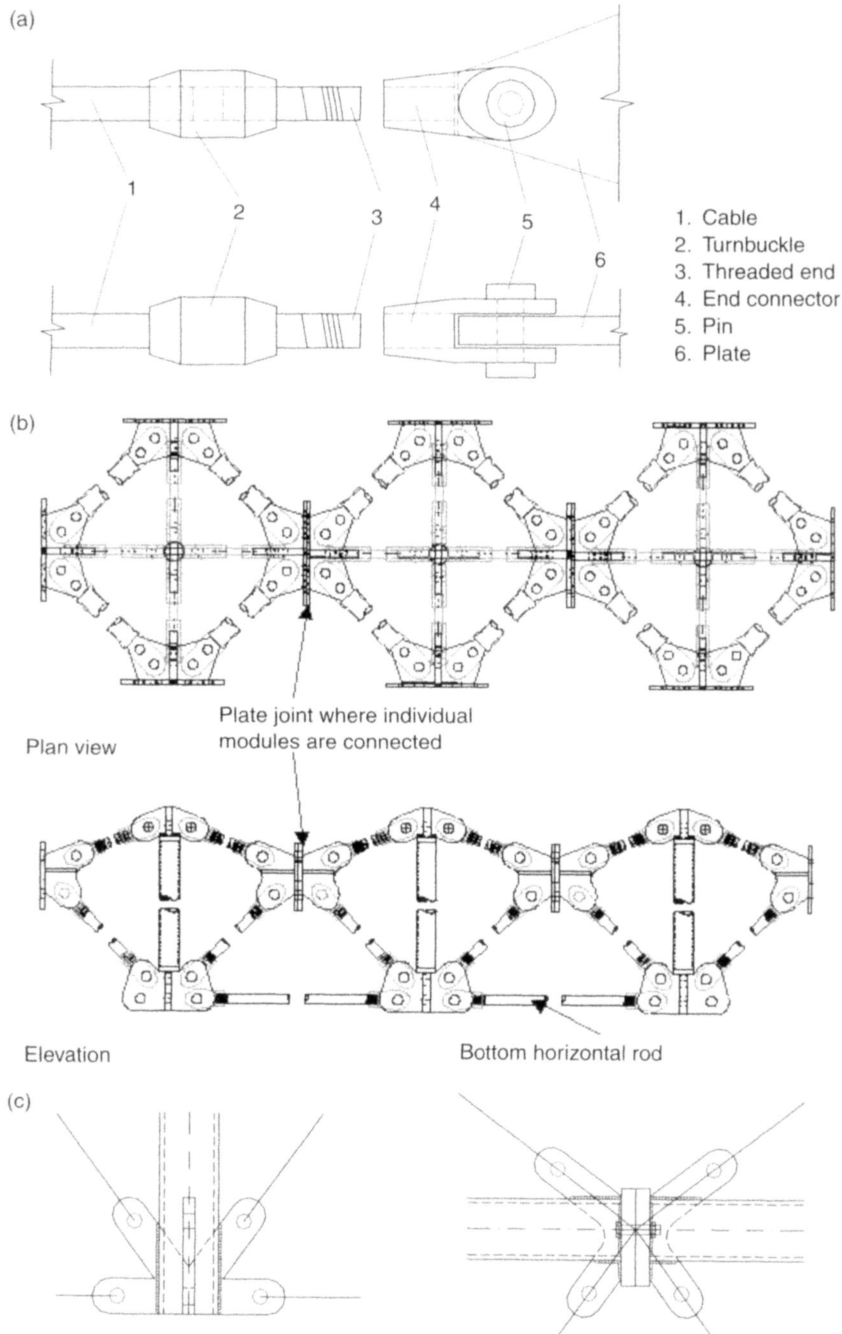

Figure 6.14 Plate joint system: (a) cable-to-plate connection; (b) RP test modules made of plate joints (source: Lee 2001); (c) improved joint design.

Alternatively, it is much cheaper to weld the ball directly to the tube (Figure 6.13(a)) by saving nuts and bolts. The typical cases are the upper and bottom layers of the RP or DP grids where only one tube is connected to the ball. It is also feasible in the compressive layer of the DP-b grid, where two half balls from adjacent simplexes can be joined easily into one as two halves always compress each other under load (Figure 6.13(b)). Meanwhile, the semi-rigidity of joints can obviate base cables in inner modules when it is preferred. In the compressive layer of the DP-a grid or RP grids, mixed welding and bolting (struts are bolted inside the ball) may be required as each half ball needs to connect two tubes and the connection of adjacent simplexes is similar to that in the DP-b grid.

It is noted that the ball can be evolved into tubular or other forms.

6.2.4 *Plate-type joint system*

Plate-type joints are popular in conventional cable structures. Each joint often contains only one plate connecting several co-planar tendons (Ishii 1999). The joint assembly contains a turnbuckle, an end connector and a pin (Figure 6.14(a)). The turnbuckle can be omitted when a tendon with reverse threads at two ends is used.

A test was done on plate-type joints designed in the RP-b grid of 8m by 8m plan (Figure 6.15) (Lee 2001). The test results comply well with the theoretical. But the 'butterfly'-shaped joints in the test model as shown in Figure 6.14(b) turn out to be too robust compared with tubes. So it is

Figure 6.15 Test model of the RP-b grid under loading.

Source: Lee (2001).

recommended that the joint be welded to the tube as one piece (Figure 6.14(c)) and high-strength material be used to reduce the size.

Plate-type joint may not be easy to be standardized like two spherical joint types. However, the detail is simple and is not sensitive to small included angles (when angle is small, just let the plate extend longer). Moreover, other components, such as flat bars, can be introduced to present new appearance (Figure 6.16).

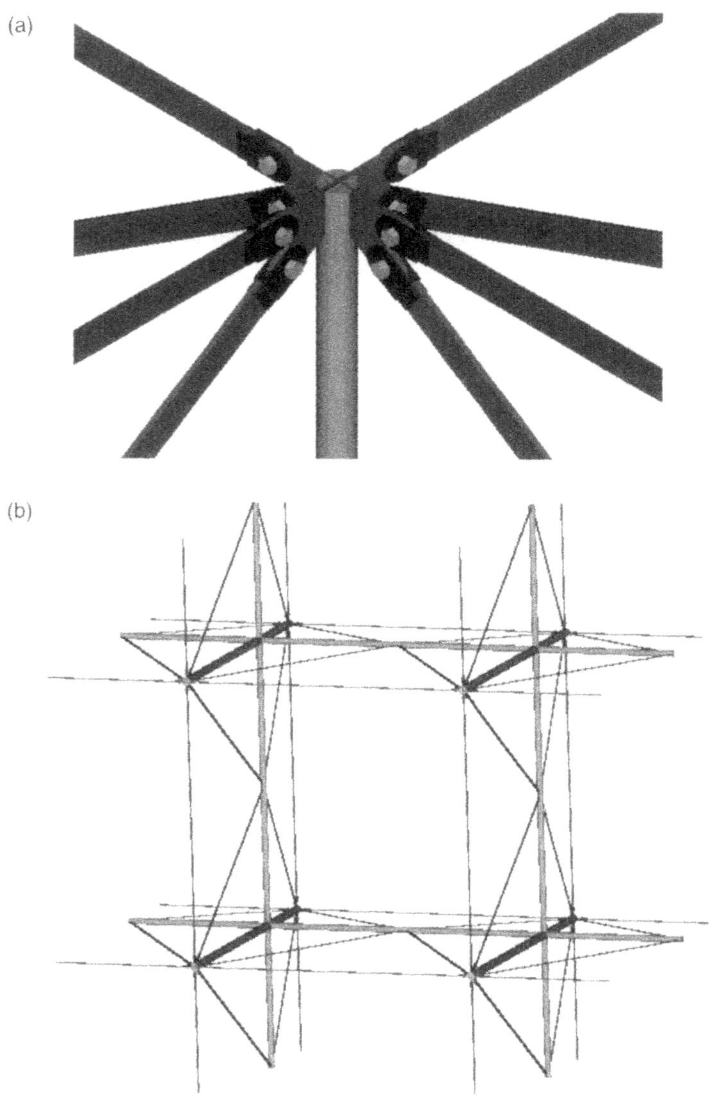

Figure 6.16 DP-b grid made of plate-type joints and flat bars: (a) top joint; (b) bottom view of the DP-b grid.

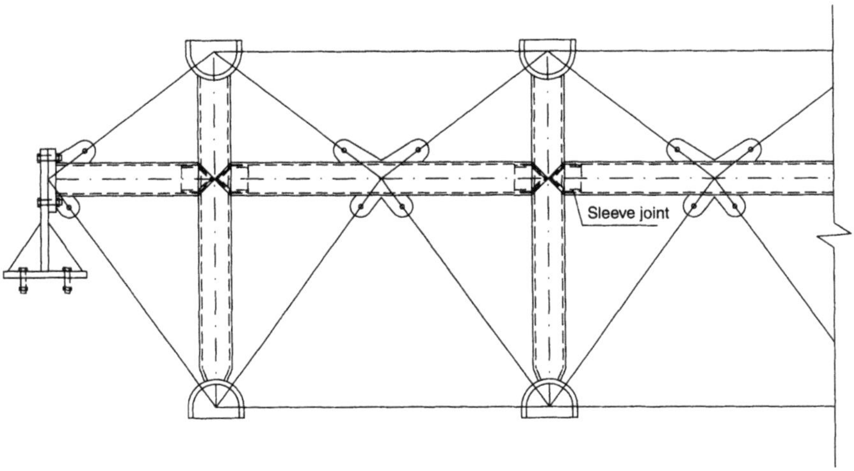

Figure 6.17 Proposal of various joints in the DP-b grid.

6.2.5 *Summary*

In cable-strut joint systems, joint design can be easier than that in conventional space frames, as tendon connection is simple and tubes are only subjected to compression, leaving more space in design. Especially inside each simplex the joint needs only to connect tubes to sustain compression, so the connection is simple, for example sleeve joint. In hollow ball joints and plate-type joints, tubes of large diameter can be designed to reduce weight without incurring component congestion. The mixed use of various joint systems may fully express their expertise, for example, plate-type and ball joints in a DP grid (Figure 6.17). However, better solutions shall be found from engineering practice.

When joint weight is considered in the structural design of cable-strut grids, larger modular length is preferred as cable-strut grids normally contain more joints than space trusses under the same modular length. Meanwhile, large modular length is beneficial for cable-strut grids of small bar lengths. As the general principle, the modular length design in the solid joint system may produce equivalent number of joints to that of space trusses, whereas the modular lengths in the hollow joint or plate-type joint system can be relatively smaller (like those given in Chapter 5).

6.3 Deployment studies of cable-strut grids

6.3.1 *Introduction*

Deployable structures can be widely used in temporary construction roofs, market roofs and temporary warehouses, etc. They are especially useful in military purposes such as field hospitals, field camps, etc. With the development of astronomical techniques, deployable space antennas and payload systems are applied to satisfy the restriction of volume and weight in space station. It is believed that antennas using primarily tensile members will play a major role in future large space antennas. But even so, the potential of application is yet to be exploited.

Different types of deployable structures have been brought forth, which can refer to Hanaor's collection (Hanaor 2000). Deploying/folding a system requires following a series of steps including the creation of 'mechanisms', that is the verification of the system compatibility in relative motion between the links, and the stabilization and the stiffening of the system in the unfolded geometry by a state of stiffness.

Components of deployable structures include plates, bars and cables, etc. In bar systems, which are used most frequently, deployment can be realized respectively by scissors elements (Figure 6.18(a)) (Escrig and Valcarel 1993, Escrig *et al.* 1996), energy-loaded elements (Figure 6.18(b)) (Fanning and Hollaway 1993, Hollaway and York 1995), telescopic elements (Figure 6.18(c)), or other complicated means. Meanwhile, activation of deployment can be realized by external energy such as fluid, electricity and manual force, etc., or internal energy stored in components.

As the composite system of bars and cables, tensegrity grids have been studied preliminarily on deployable properties, extending the application of cables into special functions. In Figure 6.19(a), Hanaor (1993) presented a small deployable model of a non-contiguous strut tensegrity grid composed of triangular simplexes. In the model, all struts are telescopic with O-ring seals through which air pressure is supplied by a bicycle bump. When bars are contracting, all cables are slack and the structure collapses into a bundle, allowing for the greatest reduction of the system volume.

Another folding method is the releasing of cables. A case of releasing individual cable is the flattening (one-way folding) of a square tensegrity truncated pyramid by lengthening a pair of opposite diagonal cables (Figure 6.19(b)) (Bouderbala *et al.* 1997). Another case is to deploy the simplex through sliding cable pairs over joints (Figure 6.19(c)), and these deployable modules are connected at site into a non-contiguous strut dome so as to avoid the requirement of a crane (Liapi 2000). When sliding groups of cables for deployment, the integral cable tension design for prestressing tensegrity grids is suitable (Figure 6.20) (Wang and Liu 1996). In the design, diagonal cables are made continuous over joints so that the grid can be contracted into a bundle at the beginning, and stretching these cables realizes the deployment of the whole grid.

Finally, the combination of releasing cables and telescoping struts is also feasible for tensegrity systems but might be complex.

6.3.2 *General principles*

Deployable cable-strut grids are made of simplexes that can concatenate in a repeatable fashion to form a desired structure. Each of these simplexes

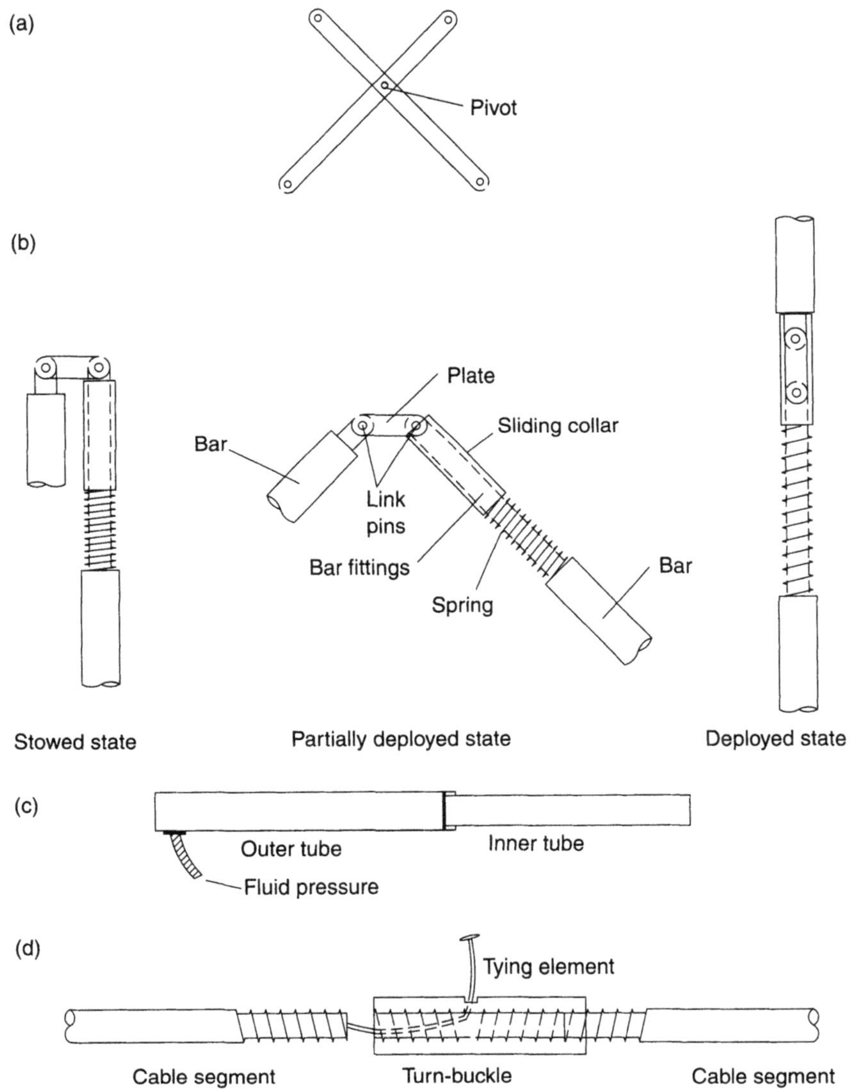

Figure 6.18 Deployment methods: (a) scissors element; (b) energy-loaded element; (c) telescopic strut; (d) releasing cable.

undergoes determinate relative motion between the links, to transform from an initially folded to finally deployed configuration, without the need for tedious assembly steps. In addition to this grid deployable method, the system may also follow module deployable method. Each deployed module is then connected into a grid (in a domical grid, the sequence shall be from the periphery to the centre). In this aspect, contiguous strut configurations

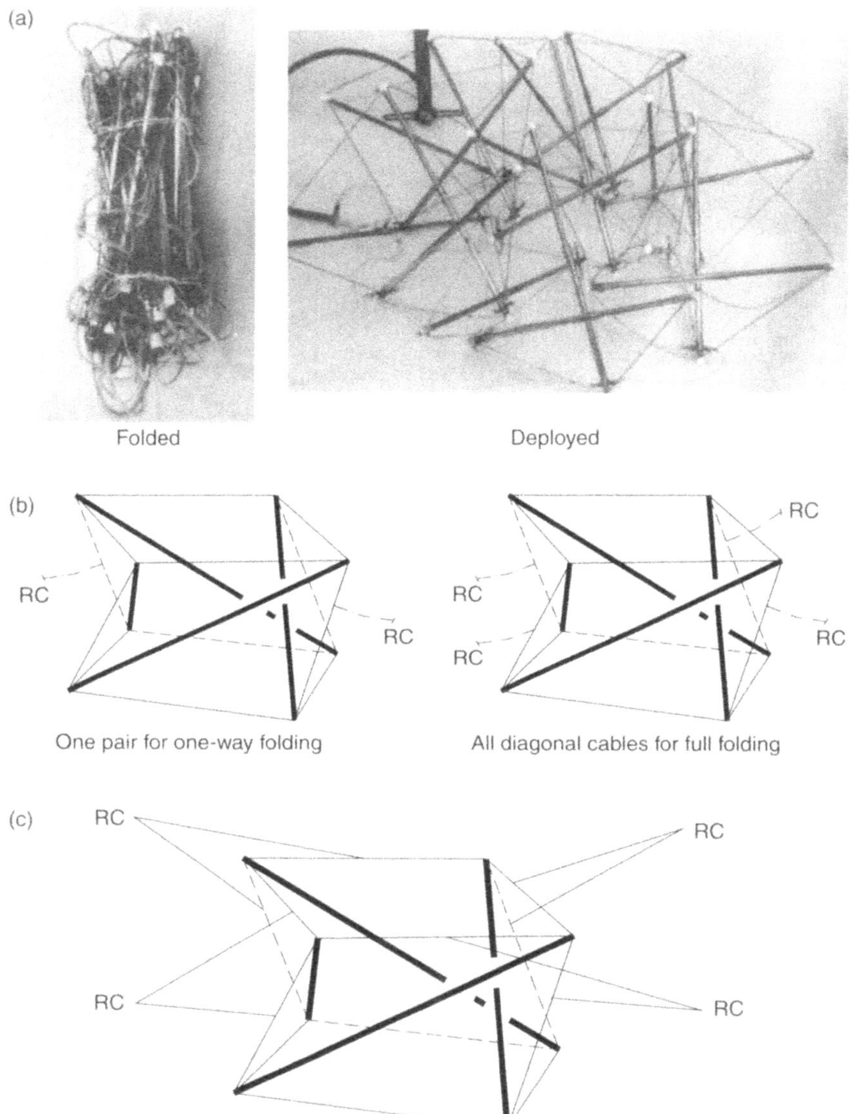

Figure 6.19 Deployable tensegrity models: (a) deployable tensegrity dome model of TS method (source: Hanaor (1997); courtesy: John Wiley & Sons); (b) releasing diagonal cables for square tensegrity prism (adapted from Bouderbala (1997)); (c) deployable tensegrity module by sliding cable pairs over joints (adapted from Liapi (2000)).

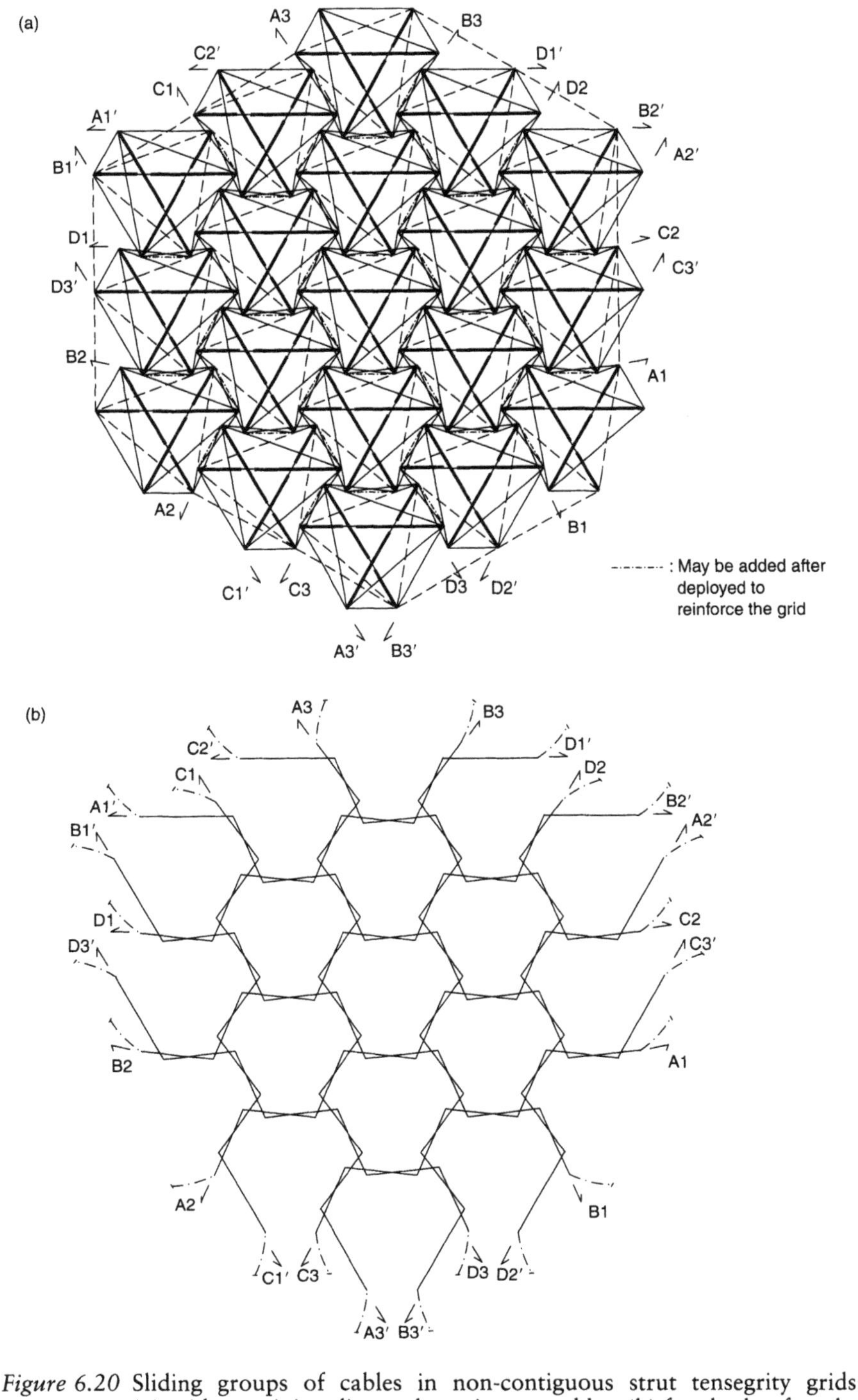

Figure 6.20 Sliding groups of cables in non-contiguous strut tensegrity grids: (a) grid containing diagonal continuous cables; (b) free-body of twelve diagonal continuous cables for prestressing.

Source: Adapted from Wang and Liu (1996).

Type 'b' for various cable-strut grids is preferred. It is generally noted that non-contiguous strut configurations have advantages in simplified joints among simplexes and that the geometrically flexible forms have merits in reduced elements. These merits in deployable functions are dominant when the requirement of stiffness is not restricted and load is small.

In tensegrity grids, telescoping struts and releasing cables are studied as deployment methods (actually, elongating struts and shortening cables are also two methods for applying prestress). In cable-strut grids, energy-loaded struts is also a choice. During the deployment process, elements in simplexes or grids can be classified into three types, including constant elements (CEs), slackening cables (SCs) and activating elements. The length of CEs is constant during deploying process neglecting the deformation due to inner forces, and such elements can be either struts or cables. SCs slacken at the folded state and during deployment process and finally, become taut at the deployed state. In the figures of this section, SCs are represented by dotted lines or omitted for simplicity.

The length of an activating element is variable so as to realize the deployment. Generally, activating elements can be classified into three types in the present study depending on which deployment method is used.

- *Telescopic struts (TSs).* A telescopic strut is activated by fluid pressure when the magnitude of elongation is relatively large (Figure 6.18(c)). It requires an external energy supply system but the grid can be deployed simultaneously. Another method is turning screw, which is free of energy supply and is normally used for small magnitude of elongation. However, the magnitude can be increased significantly by attaching a pin to lock two strut segments before fastening them through turning screw (Figure 6.21).
- *Energy-loaded struts (ESs).* They are deployed when energy is released, and folded when energy is restored (Figure 6.18(b)). The mechanism operates in such a way that when it deploys, hinge that connects two strut segments locks, and then the two behave as a single continuous piece.
- *Releasing cables (RCs).* Cables can be released individually or continuously through joints. Where released individually, folding of the simplex is realized through lengthening cables and deploying of the simplex through restoring cables to the structural length. Conventionally, releasing cables is realized by a pulley. A method of realizing RCs recommended here is to attach a tying element (e.g. a rubber band) through the turnbuckle to a cable segment (Figure 6.18(d)). When released continuously through joints, cable pairs or cable groups are continuous over the joint so as to realize the collapse of the unit (Figure 6.19(c)) or the grid (Figure 6.20). Cable pairs are fixed after deployed but cable groups are not fixed if the grid is of low frequency.

The combination of these deployment methods may be feasible but is out of the scope. In addition, a special case that there is no change in component

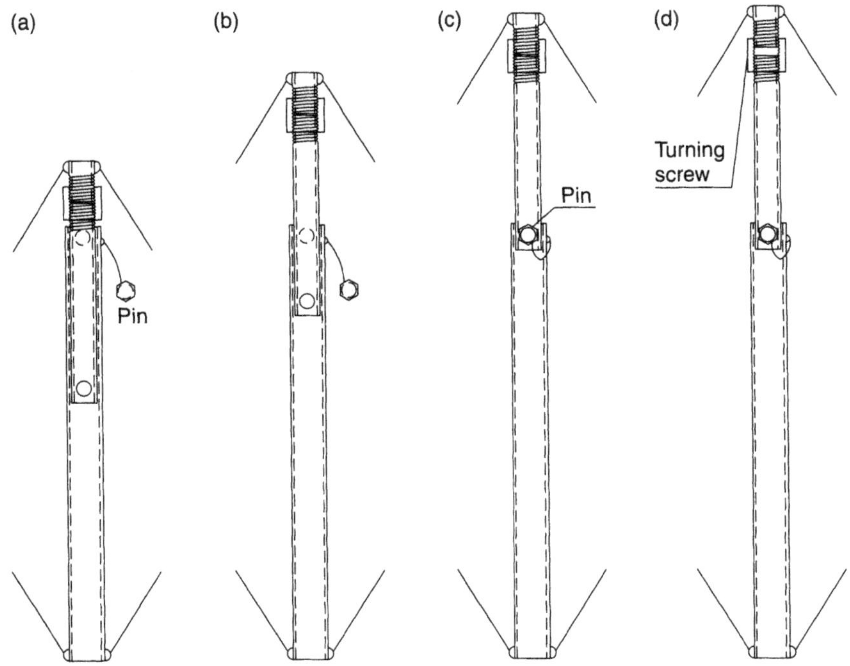

Figure 6.21 A telescopic strut locked by a pin: (a) stowed; (b) lifting; (c) locked; (d) fastened.

length is presented in Figure 6.22 for the DP-b configuration made of the square-based simplex. In each simplex, one pair of collinear horizontal struts is fixed to the vertical struts. One-way folding is realized by rotating the other pair of horizontal struts and the coplanar edge cables. The pair of struts is joined as one piece and locked to prevent rotation after deployed in position. All cables in the base are removed and additional bars are added to the edge for stability after deployed.

6.3.3 Telescoping strut method

Most cable-strut grids can be deployed by telescoping bars except that RP grids are not so convenient. Like the case in tensegrity simplexes, APs, ATPs and Ps can be deployed by telescoping either lower or upper inclined struts. However, DP grids, ATP2 grids, AP2 grids and P2 grids can be deployed more conveniently as only one telescopic strut (i.e. the vertical strut) is required in each simplex, therefore, the deployment process is simplified significantly.

In a DP, the upper and lower vertical struts can work together to form 'one' telescopic strut (Figure 6.23(a)). As the lower vertical strut is usually longer and thus of larger outer diameter, it is designed as the outer tube segment, and

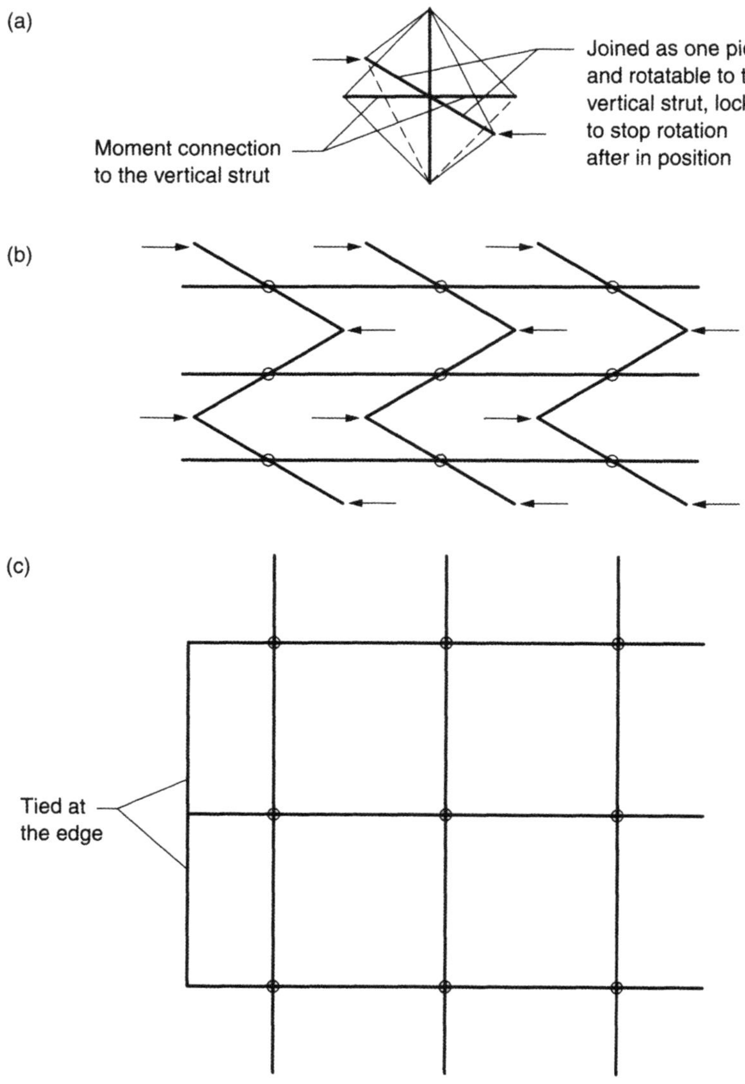

Figure 6.22 One-way folding of the DP-b grid composed of the square simplex with constant component length: (a) a deployable DP simplex; (b) intermediate state; (c) deployed state.

the upper strut becomes the inner tube segment. The procedure of deploying a triangular DP is illustrated in Figure 6.23(b)–(d). Three horizontal struts and three upper edge cables are CEs whereas other cables are SCs. For the convenience of folding a slack cable, an articulated joint may be introduced. It is interesting to note that the behaviour of a DP resembles an umbrella.

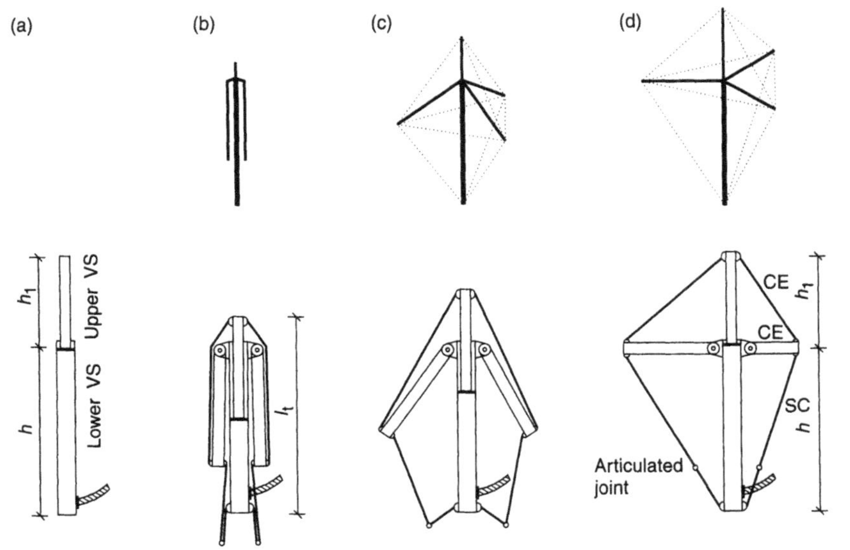

Figure 6.23 Deployment process of a DP: (a) two VSs as a telescopic strut; (b) stowed state; (c) intermediate state; (d) deployed state.

In a P2, representing an ATP2 or AP2, deployment is realized by telescoping the vertical strut, which is illustrated in Figure 6.24. The design of the vertical strut in the figure takes into consideration the reduction of the height of the simplex. Eight inclined struts and four vertical edge cables are CEs. Other cables are SCs.

The deployment should sometimes satisfy geometrical relations in order to make sure that the system can be fully folded. In the DP simplex, for example, the relation is simple as expressed by the following equation:

$$h - h_1 \geq g_{\min} \tag{6.5}$$

where h, h_1 are the length of the lower, upper vertical strut, respectively, $g_{\min}$ is the minimum gap required between two struts.

The deployment process of a DP grid composed of the triangular simplex is illustrated in Figure 6.25. In the figure, upper and bottom cables are not shown. The grid is illustrated as flat form representing the domical and cylindrical forms in actual application. In addition, square (or rectangular) simplex can be one-directionally folded to form expandable towers or arches.

6.3.4 *Energy-loaded strut method*

Energy-loaded strut method can be used to realize one-way or full folding of a simplex. Reversing the folding process realizes the deployment. An RP

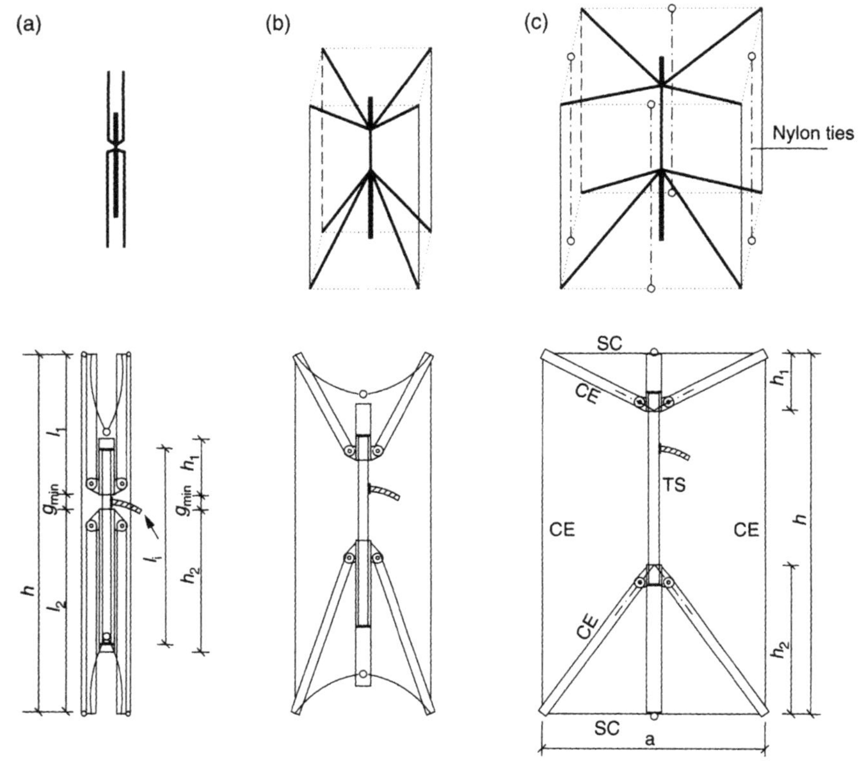

Figure 6.24 Deployment process of a P2 (stabilizing cables omitted): (a) stowed state; (b) intermediate state; (c) deployed state.

Source: Wang and Li (2003b).

can be one-way folded by folding horizontal struts in the strut plane (Figure 6.26), then deployed following reverse steps. All edge cables slacken.

A DP can be fully folded by two methods. One is to fold horizontal struts (Figure 6.27(a)). The other is much simpler, just to fold the upper vertical strut (Figure 6.27(b)), for which an additional fastening nut may be required.

An ATP, AP or P may be fully folded by folding upper inclined struts (Figure 6.28(a)). In addition, it is easy to realize the one-way folding of an ATP2, AP2 or P2 by folding only the vertical strut (Figure 6.28(b) and (c)). In a triangular P2, for example, a pair of upper or bottom struts rotate in the plane shared by the connected slackening base cable (Figure 6.28(b)), and meanwhile, the vertical strut is folded towards the opposite vertical edge cable. In a square P2, the vertical strut has two folding directions, corresponding to two grid types respectively (Figure 6.28(c)).

Note that energy-loaded joints may also be applied to connect simplexes, for example, the case in Figure 6.22. Further investigation is needed.

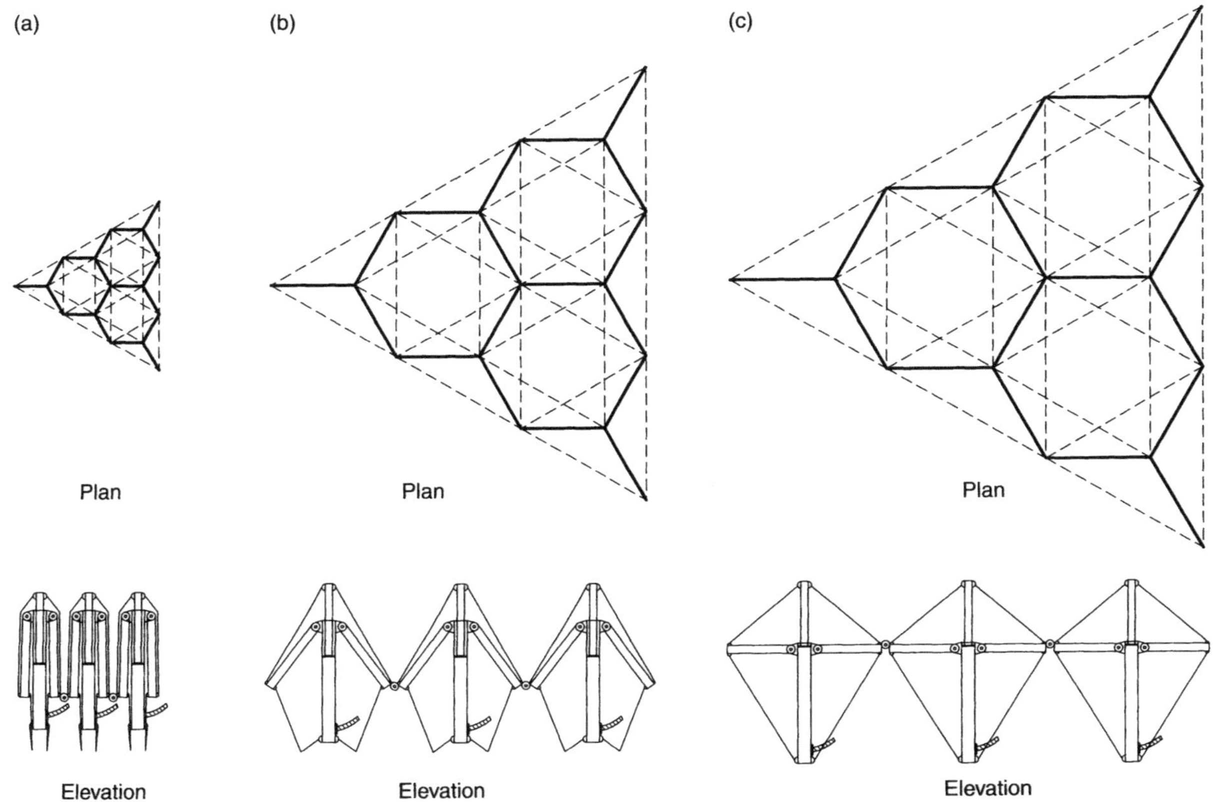

Figure 6.25 Deployment process of the DP grid: (a) stowed state; (b) intermediate state; (c) deployed state.

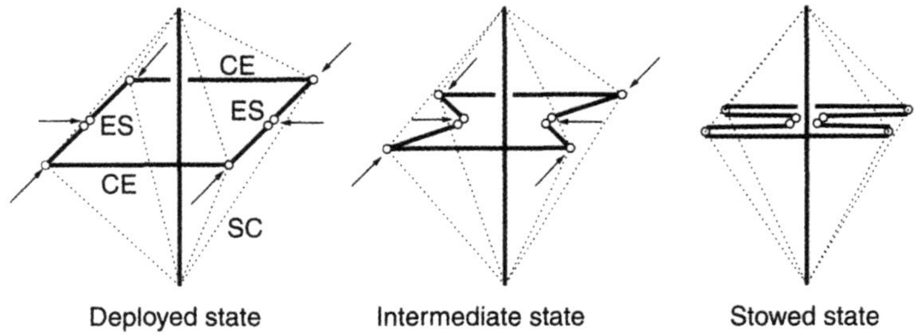

Figure 6.26 Folding a RP by ES method.

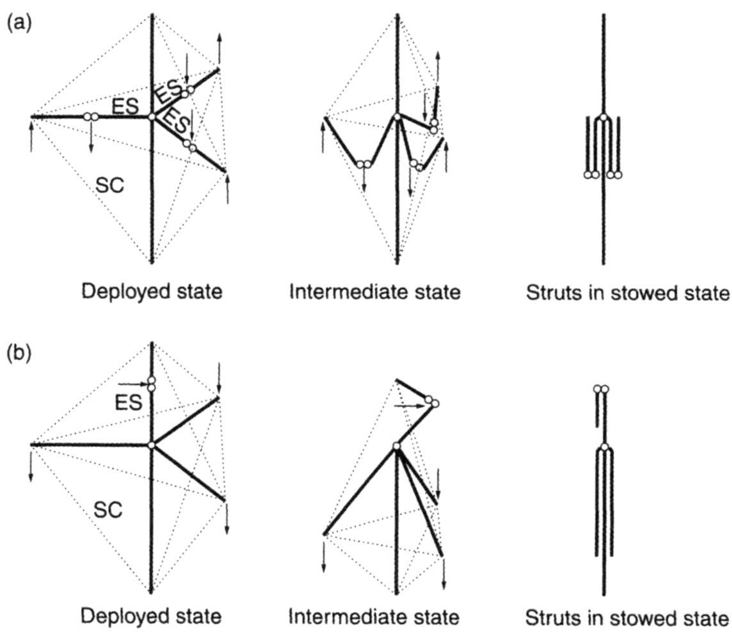

Figure 6.27 Folding a DP by ES method: (a) folding horizontal struts; (b) folding the vertical strut.

6.3.5 Releasing cable method

Releasing cable method is relatively cheaper than other methods but normally at the expense of longer operation time.

Releasing individual cables

When releasing individual cables for DPs, for example, upper inclined cables are lengthened to allow horizontal struts rotate downwardly in the

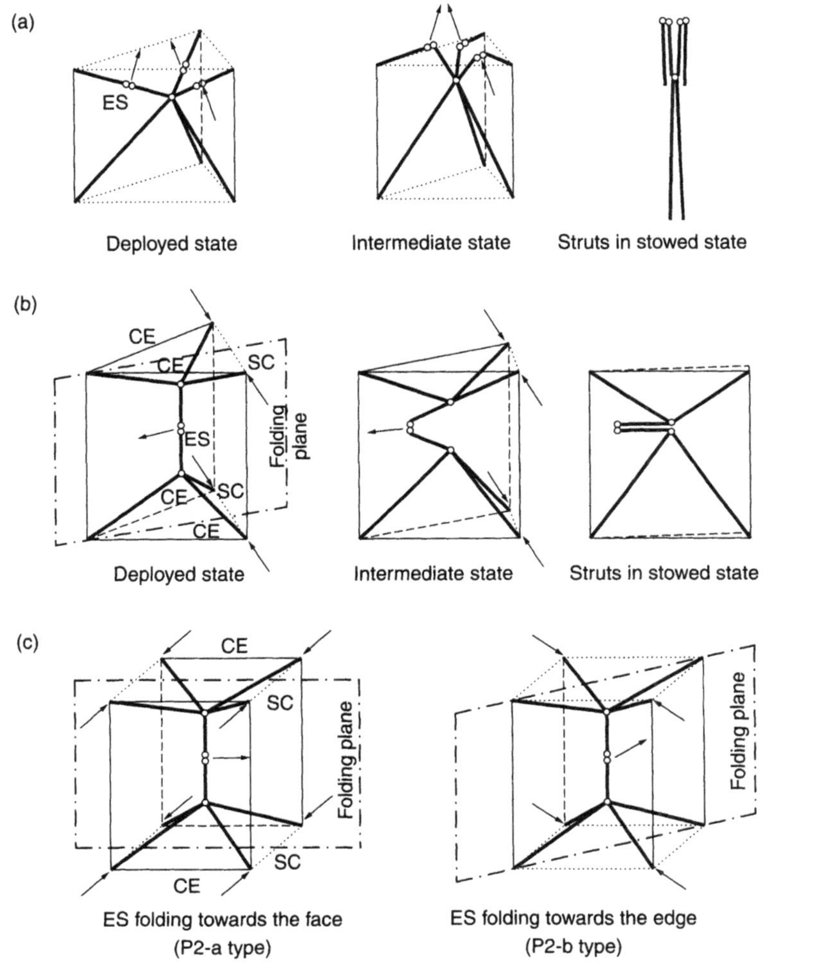

Figure 6.28 Folding P simplexes by ES method: (a) full folding of a P; (b) one-way folding of a triangular P2 (ES collapsing towards the edge); (c) one-way folding of a square P2.

Source: Wang and Li (2003b).

vertical plane (Figure 6.29(a)). The simplex is fully stowed like an umbrella. As a case of one-way folding, upper and bottom cables in the DP-a grid along the folding plane are RCs (Figure 6.29(a)).

Releasing individual cables is also convenient in the grid composed of triangular or square Ps as lateral bracing cables are not required. Therefore, the triangular or square P can be fully stowed only by lengthening vertical cables, which is illustrated in Figure 6.30(a).

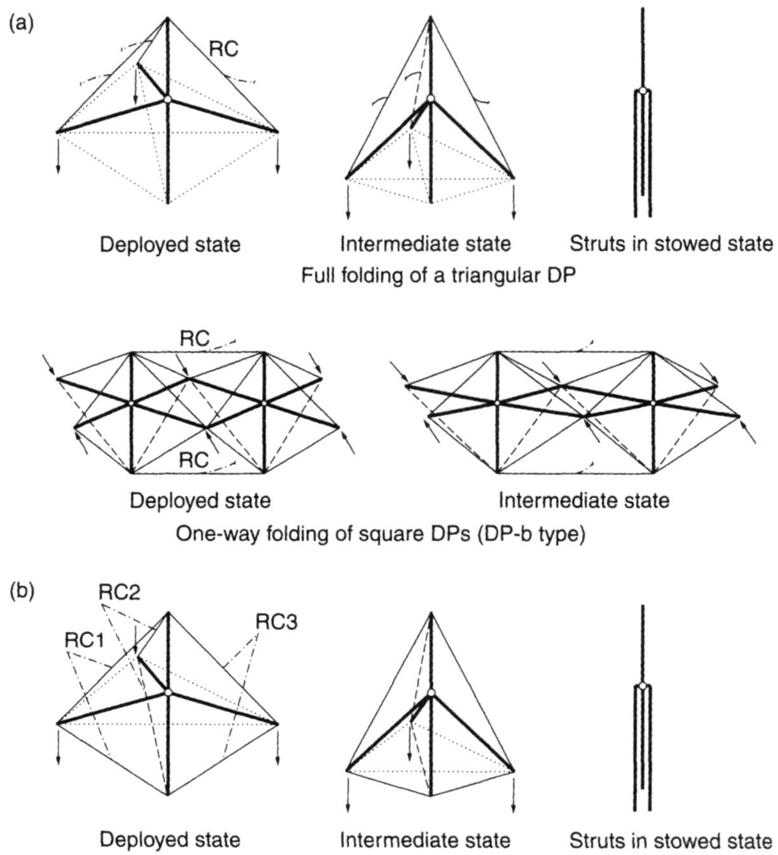

Figure 6.29 Folding DPs by releasing cable method: (a) folding DPs by individual cables; (b) folding of a DP by sliding cables.

Sliding cable method

When two cables are continuous over joints, the sliding cable method applied to tensegrity simplexes is also applicable to the grid composed of triangular or square Ps with higher structural efficiency and more stable geometry. As shown in Figure 6.30(b), an upper cable and a vertical cable form a group to slide over the common joint, and bars always rotate in the vertical plane. Geometrical relation should be satisfied, that is, the total length of two cables being greater than that of two bars (one upper, one lower). In the square prism, as an example,

$$h_1 + h_2 + a \geq \sqrt{h_1^2 + a^2/2} + \sqrt{h_2^2 + a^2/2} \tag{6.6}$$

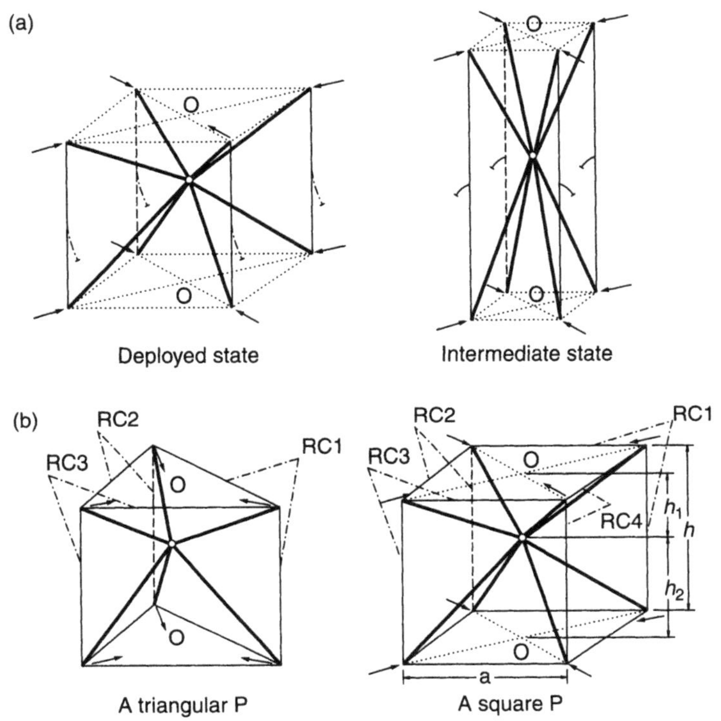

Figure 6.30 Folding Ps by releasing cable method: (a) folding Ps by releasing individual cables (full folding of a square P); (b) folding Ps by sliding cables.

Source: Wang and Li (2003b).

The definition of parameters refers to Figure 6.28(b). It is easily satisfied so long as the depth is not too small. For example, when $h_1 = h_2 = h/2$, it is just required $h > a/2$.

The application of sliding cable pairs to a DP is more straightforward. Each pair of upper and bottom inclined cables forms an RC during the deploying/folding process (Figure 6.29(b)). The spirit can be applied to other simplexes.

When sliding cable groups, the method for tensegrity grids (Figure 6.20) is clearly suitable for the AP or ATP grid with more stable geometry.

6.3.6 *Summary*

Generally, cable-strut grids are lightweight and can achieve much less stowed volume and decrease the number of joints that have complicated mechanisms. They are superior to conventional bar systems when difficulties are caused by increased weight, packing volume and number of joints.

As it is a design requirement that deployable grids are restrictedly free-standing, CP grids are only module deployable. The versatility of other cable-strut grids, including non-contiguous strut forms, in deployment functions is expressed by telescopic strut method, energy-loaded strut method and releasing cable method, etc. and even constant component length (Figure 6.22). Discussions show that deployment of some cable-strut grids can be realized by telescoping only one strut in each simplex and many choices are provided by energy-loaded method and releasing cable method (releasing individual cable, cable pair/group) in one-way or full folding. However, the discussions are not sufficient to cover all possible cases.

According to the special functions and requirements, various types of deployable structures may be developed respectively into market. As cable-strut grids are pin-jointed systems, mechanical analysis during the folding/deploying process seems unnecessary unlike the case of scissor-element systems (Figure 6.18(a)). Further studies include the selection of deployable methods, energy supply design, joint design, geometrical study in consideration of the actual dimensions of components and joints, form-finding

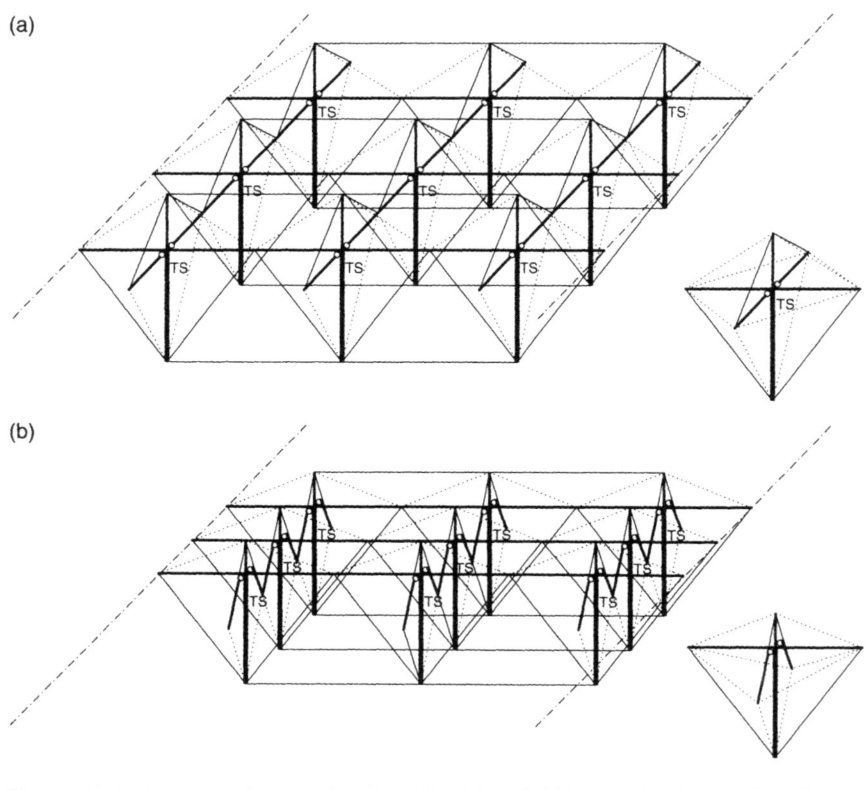

Figure 6.31 Concept of retracting the DP grid by folding method: (a) original state; (b) partially retracted state.

of various geometry and cladding, and dynamic behaviour, etc. Finally, experiments are required to test the efficiency of these types.

Another important special application of cable-strut grids lies in retractability. Retractable structures meet people's desire to the nature and thus the study becomes popular. The more conventional way of retraction is realized by the movement of grid panels (the grid is divided into several independent panels) along rails (Ishii 2000). In this aspect, the advantage of cable-strut grids in lightness may facilitate the process. Another way of retracting is realized through folding cable-strut grids simultaneously. As an example, one-way folding of the DP grid by the TS method is illustrated in Figure 6.31. More forms can be developed with the accumulation of engineering practice.

7 Architectural aspect of cable-strut systems

Lightweight structural systems are becoming increasingly valuable for the architecture and design fields. Tensegrity systems, and extensively, cable-strut systems have the potential of offering interesting viable solutions for architecture. Their free-standing nature is the additional merit. Structural forms in Chapter 4 for cable-strut grids, especially non-contiguous strut ones, shall be also innovative in architecture but are considered 'basic' forms. This chapter explores the architectural aspect developed from basic forms.

As cable-strut systems are developed from tensegrity systems, the chapter starts from the brief review of the contributions of tensegrity concept in architecture. Then various dimensional cable-strut systems are proposed accordingly. Moreover, a wide variety of new architectural cable-strut forms can be developed from special cable-strut configurations, simplexes with ornamental supplements, innovative roof shape and roof sculpture design with matched structural design, and reshaping cable-strut simplexes themselves etc. Finally, the principle of forming higher cable-strut polyhedra is illustrated.

In addition to the innovation in forms, new architectural art can be presented by the employment of new material, which is beyond the scope. One example, struts in cable-strut simplexes or grids can be made of timber components. The timber member normally needs an end cap, for example, a joint cone (Imai *et al.* 2002) to be connected to a node, but the property that the member is only subjected to compression simplifies the application significantly. Another recommendation is the use of air-inflated plastic pipe.

7.1 Review of architectural tensegrity forms

Tensegrity systems have natural expertise in architecture. Certain attractiveness lies in that only independent struts catch eyes from afar and/or that a wide variety of unique forms can be developed.

7.1.1 Tensegrity walls

Basic structural forms in Chapter 3 spanning horizontally are pleasing in architecture themselves. They can be developed in three dimensional ways

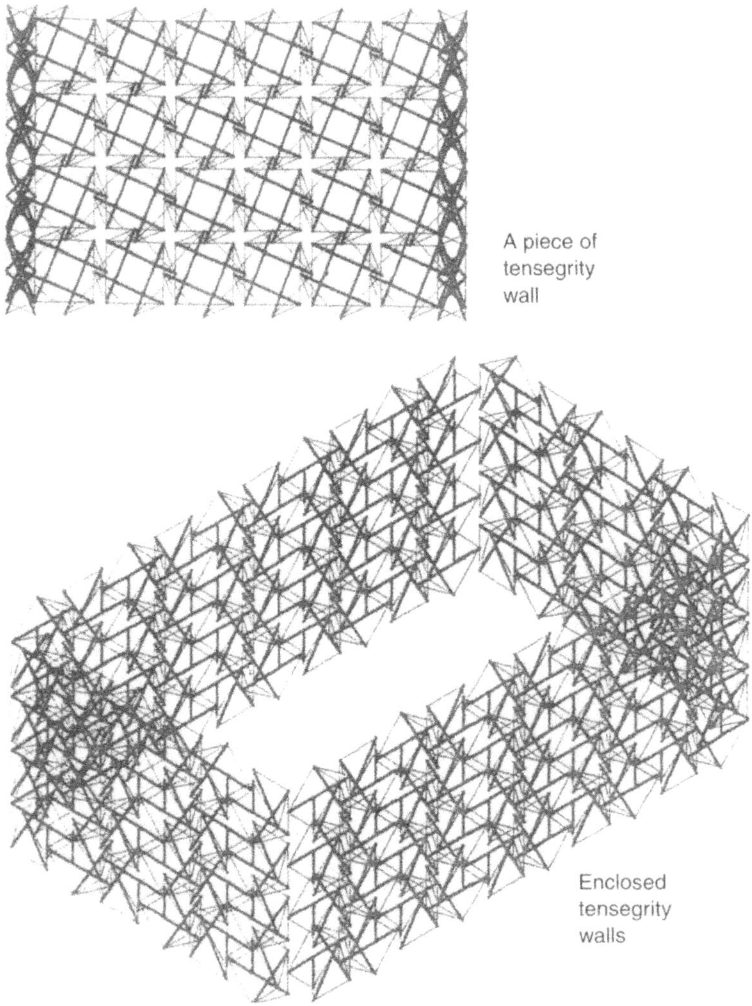

Figure 7.1 Tensegrity walls.

to extend their application, for example, in the vertical cladding of glass as tensegrity walls (Figure 7.1). Some openings shall be designed but are not shown. As normally the span of walls is not much, the effect of structural efficiency is not significant.

7.1.2 *Some plane-filling and multi-layer filling forms*

In addition to basic tensegrity forms, more architectural forms are developed including Snelson's sculptures (one-way array of X modules) and tensegrity polyhedra presented in Chapter 2. Emmerich's plane-filling forms and hyper-polyhedra, and Emmerich's and Grip's multi-layer space-filling forms are presented in this section.

From literature, Emmerich was the first to conceive double-layer non-contiguous strut tensegrity grids. His deck-like form (Figure 7.2(a)) is composed

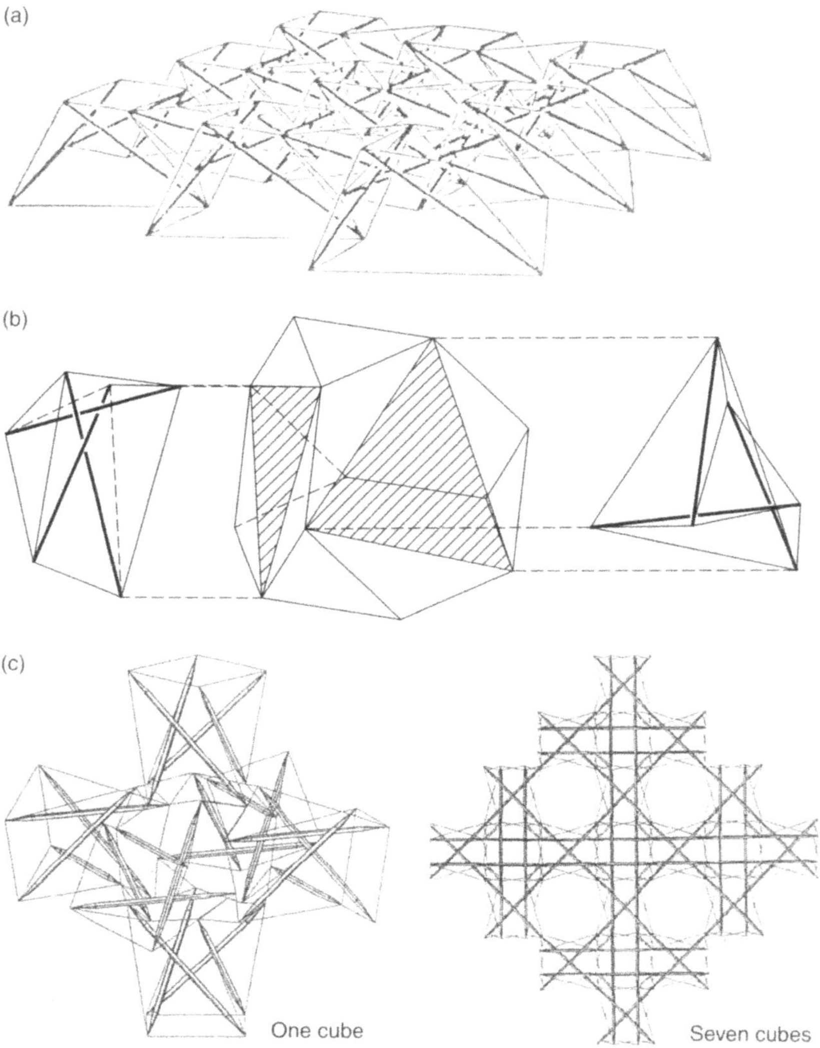

Figure 7.2 Some ornamental tensegrity forms: (a) double-layer deck-like flat space frame; (b) hyper-truncated tetrahedron; (c) tensegrity cube array.

Sources:
a Emmerich (1990); Courtesy: Multi-Science Publishing.
b Adapted from Emmerich (1990).
c Grip (1992); Courtesy: Multi-Science Publishing.

from interlaced quadralic self-tensioning simplexes having automorphic tessellation as basic pattern (Emmerich 1990). However, simplex connection seems not strong enough as only one base is connected directly. For many of other plane-filling forms, refer to the report by Hilyard and Lalvani (2000). In Emmerich's 'hyper-polyhedra', the triangular, square, or pentagonal tensegrity truncated pyramids are basic solids filled onto the faces of the polyhedron with openings (Emmerich 1990). Alternatively, simplexes can be filled to every face. These faces are hexa-, octa-, or decagonal polygons, each having vertices double those in the base of its basic solid. A hyper-truncated tetrahedron, typical of the concept, is shown in Figure 7.2(b), in which the triangular tensegrity solids are filled onto the hexagonal faces. Multi-layer space-filling forms are developed from plane-filling forms or hyper-polyhedra (Emmerich 1993).

Grip (1992) proposed a multi-layer space-filling form typical of an array of 'cubes'. Each cube is composed of six square tensegrity prisms, as shown in Figure 7.2(c). These prisms are connected in a joint-to-cable way and an empty volume is enclosed by the connected base of each simplex. To create an array of cubes, additional modules are added to the unused base polygon of each tensegrity prism, to connect a cube on each face of the original cube (Figure 7.2(c)). Thus each cube shares one tensegrity prism with its neighbouring cube.

7.1.3 *Tensegrity cable dome*

A 'special' application of tensegrity forms is that the enclosed radial linkage of (stiffened) tensegrity truncated pyramids produces a tensegrity ring beam to support a cable dome (Figure 7.3). As simplexes are connected in

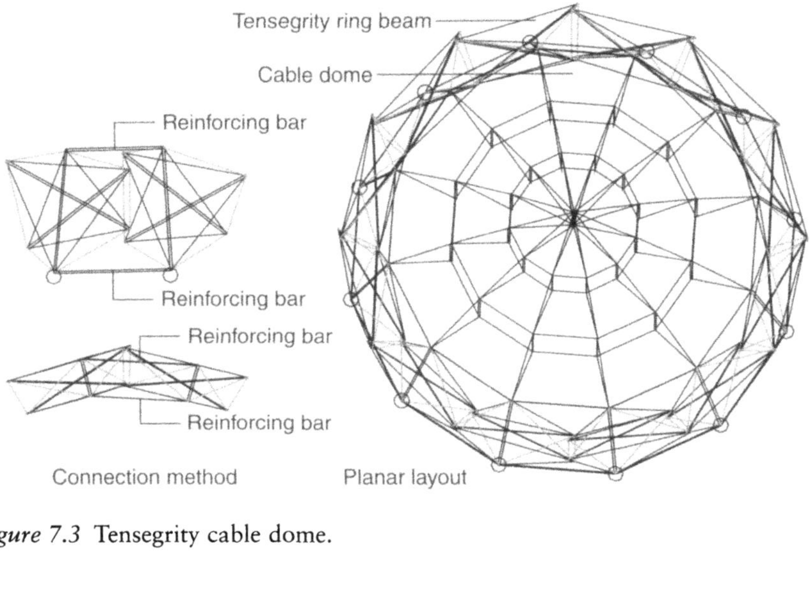

Figure 7.3 Tensegrity cable dome.

a node-to-cable way, the ring shall be reinforced by struts in order to provide sufficient stiffness (Wang 1996b). The tensegrity ring and supported cable dome as a whole conceptually form a 'pure' tensegrity model although it is unlikely to be suitable in large spans.

7.1.4 *Other tensegrity forms*

In addition, a few relatively new tensegrity forms looking different from 'basic' forms are developed recently. In Figure 7.4, the non-contiguous strut grid is basically the interlacing of triangular anti-truncated pyramids (Kono *et al.* 1999). The large base of a simplex lies in the same layer as the small base of adjacent simplexes, and vice versa. Each pair of adjacent simplexes shares a strut, whose ends connect the vertices of small bases in different layers. Besides, Motro proposed 'truss-like' forms evolved from contiguous strut tensegrity forms, which are also characteristic of long inclined struts (Motro 2002).

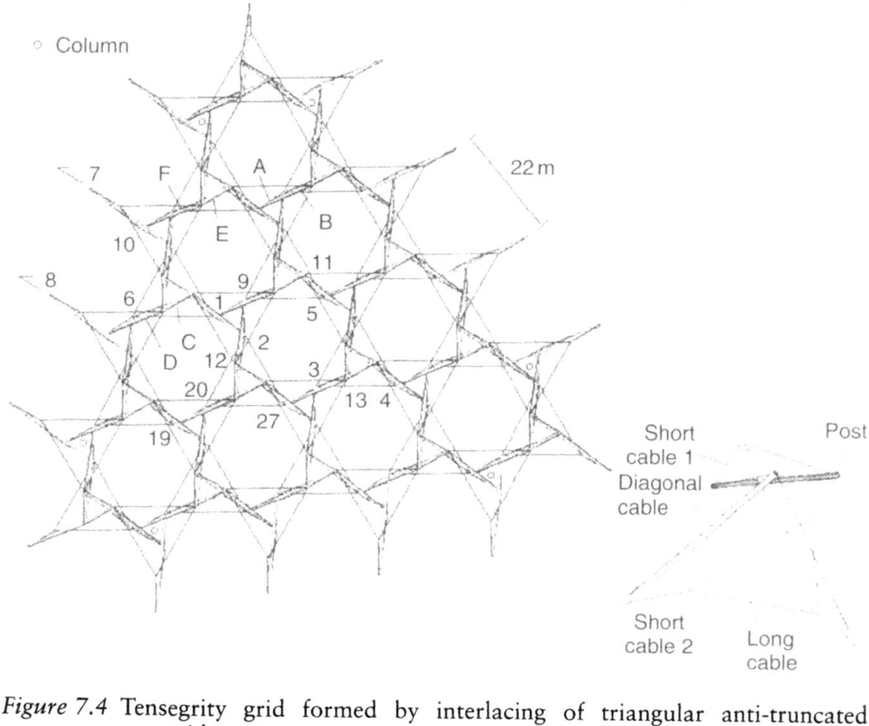

Figure 7.4 Tensegrity grid formed by interlacing of triangular anti-truncated pyramids.

Source: Kono *et al.* (1990).

7.2 Linear, plane-filling and space-filling cable-strut forms

Cable-strut simplexes are building blocks of structurally efficient load-bearing grids. Naturally, they can be used to develop more architectural forms in linear, plane-filling and space-filling forms, which are enriched from morphological studies.

7.2.1 Linear forms

Two types of linear forms are illustrated here. One is the tower form (Figure 7.5), the other the walkway (Figure 7.6). Four towers are illustrated in Figure 7.5, presenting difference appearance. Two towers in Figure 7.5(a), and (b) are composed of triangular RPs (reciprocal prisms), and triangular DPs (di-pyramids) with a CP (crystal-cell pyramids) base, respectively. The former needs four supporting points whereas the latter only three. The tower in Figure 7.5(c) is composed of triangular Ps (prisms), in which stabilizing cables are added in each simplex to improve stiffness. In addition, a tower can be designed more 'practically', for example, a P tower is designed with three legs and a CP-shaped pinnacle (Figure 7.5(d)). It is obvious that more towers can be formed from other simplexes and mixed use of some types.

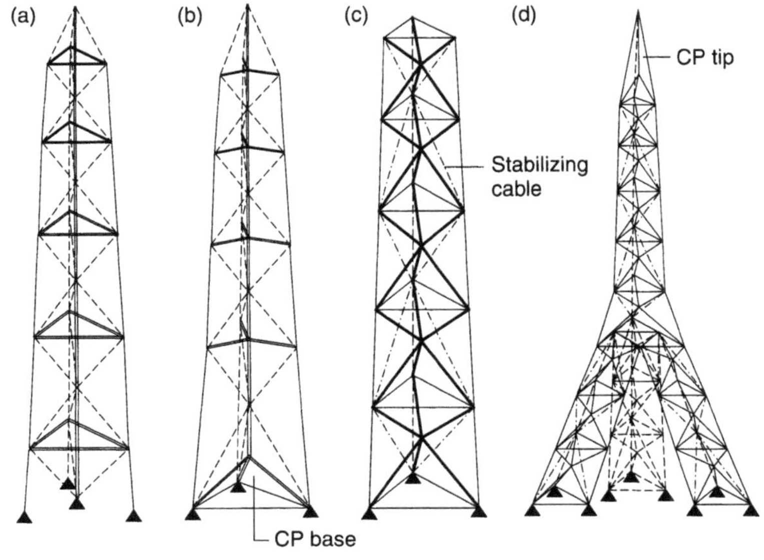

Figure 7.5 Cable-strut towers: (a) a RP tower; (b) a DP tower with a CP base; (c) a P tower; (d) a P tower with a CP tip and three legs.

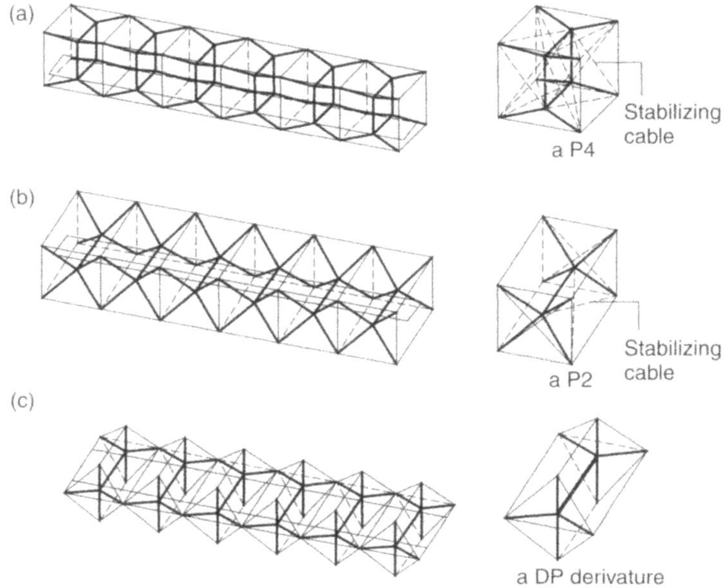

Figure 7.6 Cable-strut walkways: (a) a walkway made of P4s; (b) a walkway made of P2s; (c) a walkway made of DP derivatures.

The walkway in Figure 7.6(a) is made of square-based P4s, P simplexes with four inner joints. As shown in the figure, each simplex can be stabilized by connecting opposite vertices in each trapezoid by cables. The walkway can also be made by P2s (Figure 7.6(b)) or other simplexes. An extensive application is presented in Figure 7.6(c), in which the building block is a derivature of square-based DP (see Section 7.8).

7.2.2 *Plane-filling forms*

Most plane-filling forms built from tensegrity simplexes are also suitable for cable-strut simplexes. Some plane-filling forms are presented in Figure 7.7. In contiguous strut forms in Figure 7.7(a–d), large openings are enclosed by square simplexes, pentagonal simplexes or mixed square and hexagonal simplexes. Here, simplexes refer to all types. As shown in Figure 7.7(a), and (b), simplexes are divided into two groups, with CPs, Ps, DPs and RPs into one group and ATPs and APs into the other, but the latter is not illustrated in Figure 7.7(c) or (d).

All these forms have their corresponding non-contiguous strut forms, which are applicable where appropriate. Moreover, it is noted that a non-contiguous strut forms in Figure 7.8 composed of 'windmill' unit. In each unit, six arms of mixed pentagonal and quadrilateral simplexes are connected to a hexagon at the centre, indicating the potential.

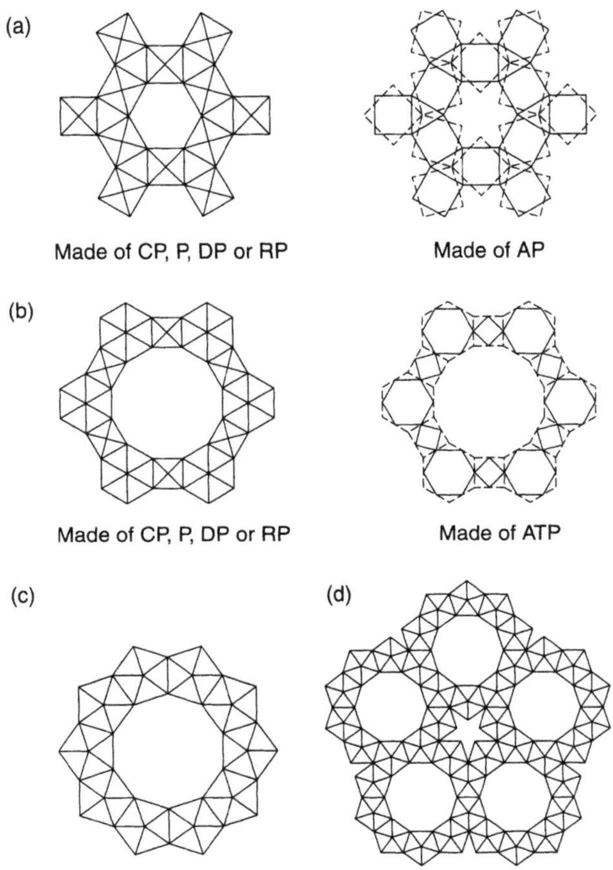

Figure 7.7 Plane-filling cable-strut forms: (a) made of square simplexes; (b) made of square and hexagonal simplexes; (c) made of pentagonal simplexes; (d) made of pentagons with small stars.

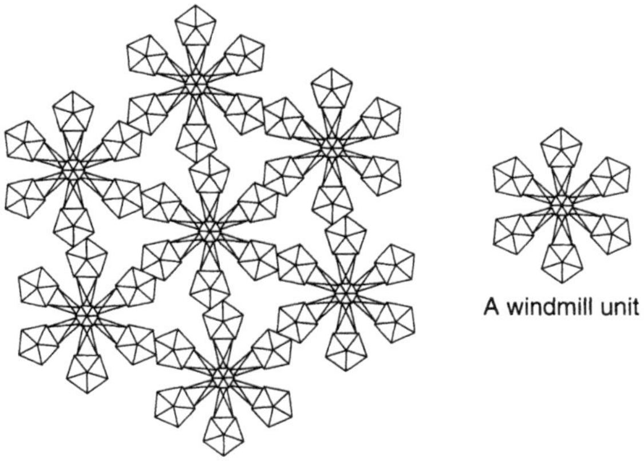

Figure 7.8 Non-contiguous strut form composed of windmill units.

7.2.3 *Space-filling forms*

Many space-filling forms for tensegrity simplexes can also be applied for cable-strut simplexes. For example, cable-strut simplexes can be inscribed into a polyhedron, including a Platonic, an Archimedian, or hyper-polyhedron, etc. as the case for tensegrity simplexes (Figure 7.2(b)). Cable-strut polyhedra can be formed based on the discussion in Section 7.8.

Plane-filling forms of cable-strut simplexes can be developed in a three-dimensional way as space-filling forms. As an example, in Figure 7.9 it is

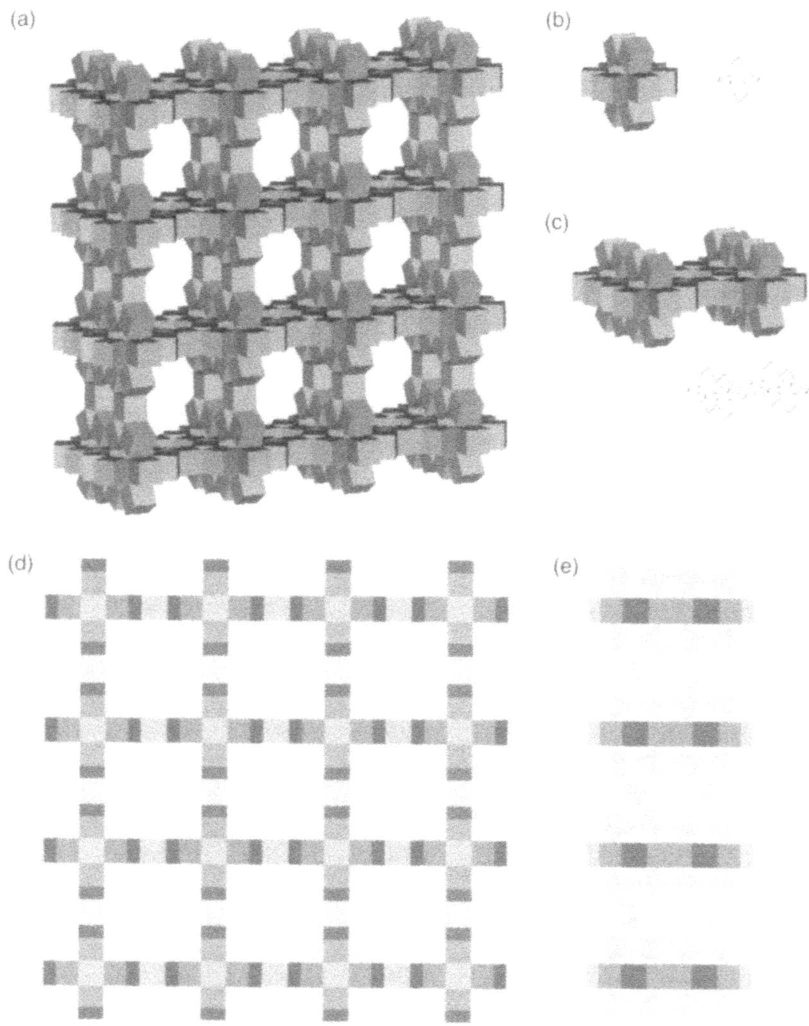

Figure 7.9 Cable-strut space-filling form composed of cubic prisms (a) isotropic view; (b) basic unit; (c) array of units; (d) front view; (e) side view.

Source: Wang and Li (2003b).

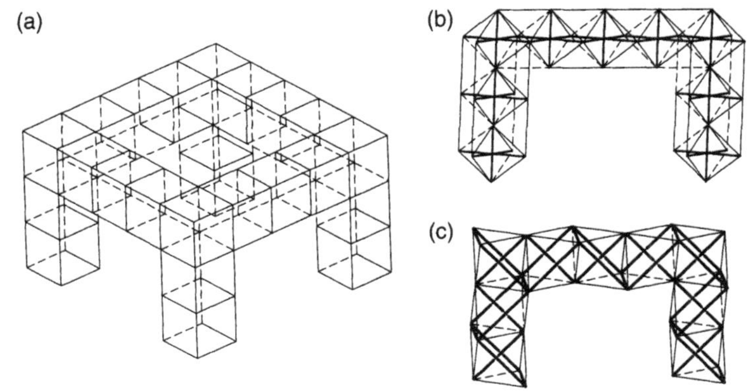

Figure 7.10 Cable-strut truss forms: (a) three-way truss; (b) truss of DP simplexes; (c) truss of RP simplexes.

illustrated that a type of space-filling form is developed from the planar form in Figure 7.7(a). Each building unit comprises of two orthogonal hexagonal rings that share two simplexes (Figure 7.9(b)). Each hexagonal ring is composed of six cubic simplexes, which can be cubic prisms (as drawn), square DPs or square RPs. Two building units are then connected by adding two simplexes to form a hexagonal ring so as to enclose empty space (Figure 7.9(c)). The array is then developed two ways (as shown in Figure 7.9(a), (d), (e) for different views) or three ways.

Another example, a three-way truss is illustrated in Figure 7.10(a), building blocks being RP simplexes (Figure 7.10(b)), DP simplexes (Figure 7.10(c)), or other cable-strut simplexes. More unexpected forms can be developed from morphological studies of polyhedra geometry. Some studies include Wester (2000), Burt (1996) and Huybers (2002), etc.

7.3 Special cable-strut configurations

Basic cable-strut simplexes can be applied into various dimensional forms based on basic configurations in which bases are always co-planar. Moreover, these simplexes can be applied into 'special' configurations that are different from these basic ones. The topic is highlighted in this section. It is interesting to note that different groups of simplexes present different connections and that the connections are different in two ways in the resulting grid. Note that these configurations can also be developed into 'spatial' forms.

7.3.1 RP and DP simplexes

RP or DP simplexes can be connected at the diagonal face (Figure 7.11). This is to say, adjacent simplexes in one way share a diagonal face, and 'vertical' struts and the base are inclined. In the other way, the struts in the base form continuous flow presenting truss-like appearance. Note that additional

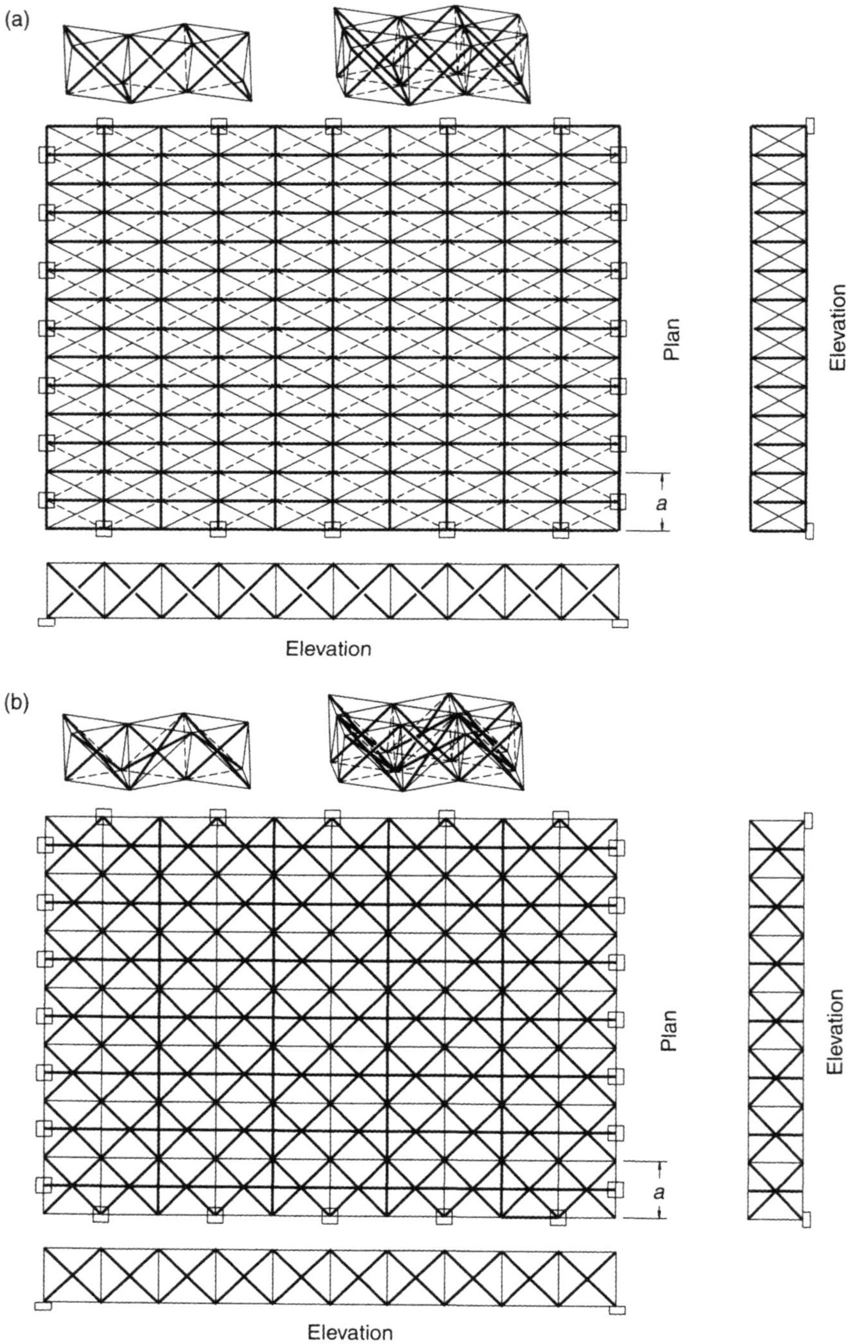

Figure 7.11 Special RP and DP configurations by connecting diagonal faces: (a) RP grid; (b) DP grid.

connecting cables are not required. In the figure, only square-based simplexes are used, but simplexes of other bases also can work.

From geometry, RP or DP simplexes can also be connected through the 'vertical' struts (Figure 7.12). Hence in one way, 'vertical' struts become horizontal and geometrically continuous in the grid. In the other way, the struts in the base are connected into an 'independent truss'. When square simplexes are applied, it is an array of squares in the RP grid (Figure 7.12(a)), and an array of crossings in the DP grid (Figure 7.12(b)) (when triangular simplexes are used, additional connecting cables are needed). The connections among these trusses are diagonal cables, horizontal struts and additional connecting cables.

7.3.2 ATP, AP and P simplexes

ATP and AP simplexes can be connected at the base (Figure 7.13). Therefore, adjacent simplexes share the base face in one way. Square-based ATP and triangle-based AP are used as building blocks in the figure, representing other simplexes. In the other way, each pair of square-based ATPs share a lateral face (Figure 7.13(a)) whereas each pair of triangle-based APs just share an edge (Figure 7.13(b)). The appearance of Figure 7.13(a) looks interesting.

In P simplexes, triangular Ps can be connected at the base to form a truss. In the grid form (Figure 7.14(a)), the grid also follows truss-like behaviour under uplift load. Square Ps are special, as lateral faces and two bases are the same, presenting the same design as basic forms. However, square P2s can be employed to present different appearance (Figure 7.14(b)).

7.3.3 CP simplexes

The triangular CP can be connected along the edge into a 'CP truss' (Figure 7.15(a)). In the CP truss, only one edge cable from each simplex is placed at the top layer, connected with adjacent ones. Two inclined struts in each simplex are placed in the upper compressive layer. It is lightweight and can be used in one-way spanning under downward load only. The grid can take uplift force with lower inclined struts in the compressive layer when it is two-way supported. Two-way grid forms can also be realized by the orthogonal array of the trusses, the basic unit is shown in Figure 7.15(b).

7.4 Special structural and roof design

Examples below show some 'special' roof forms with matched structural design, which are different from 'normal' simplex or grid application. The flexibility in the design shows the potential application of cable-strut systems in structural design and architecture.

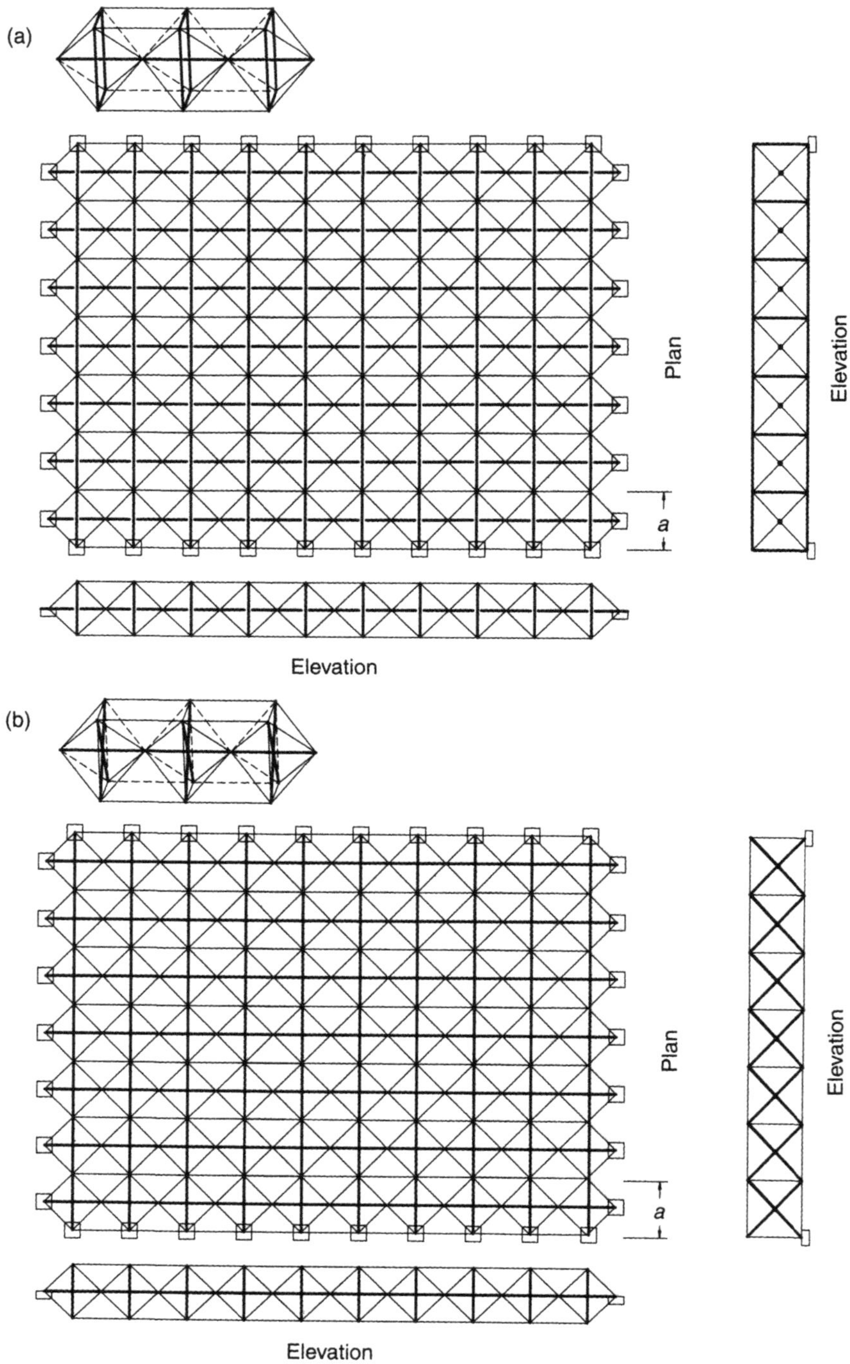

Figure 7.12 Special RP and DP configurations by connecting vertical struts: (a) RP grid; (b) DP grid.

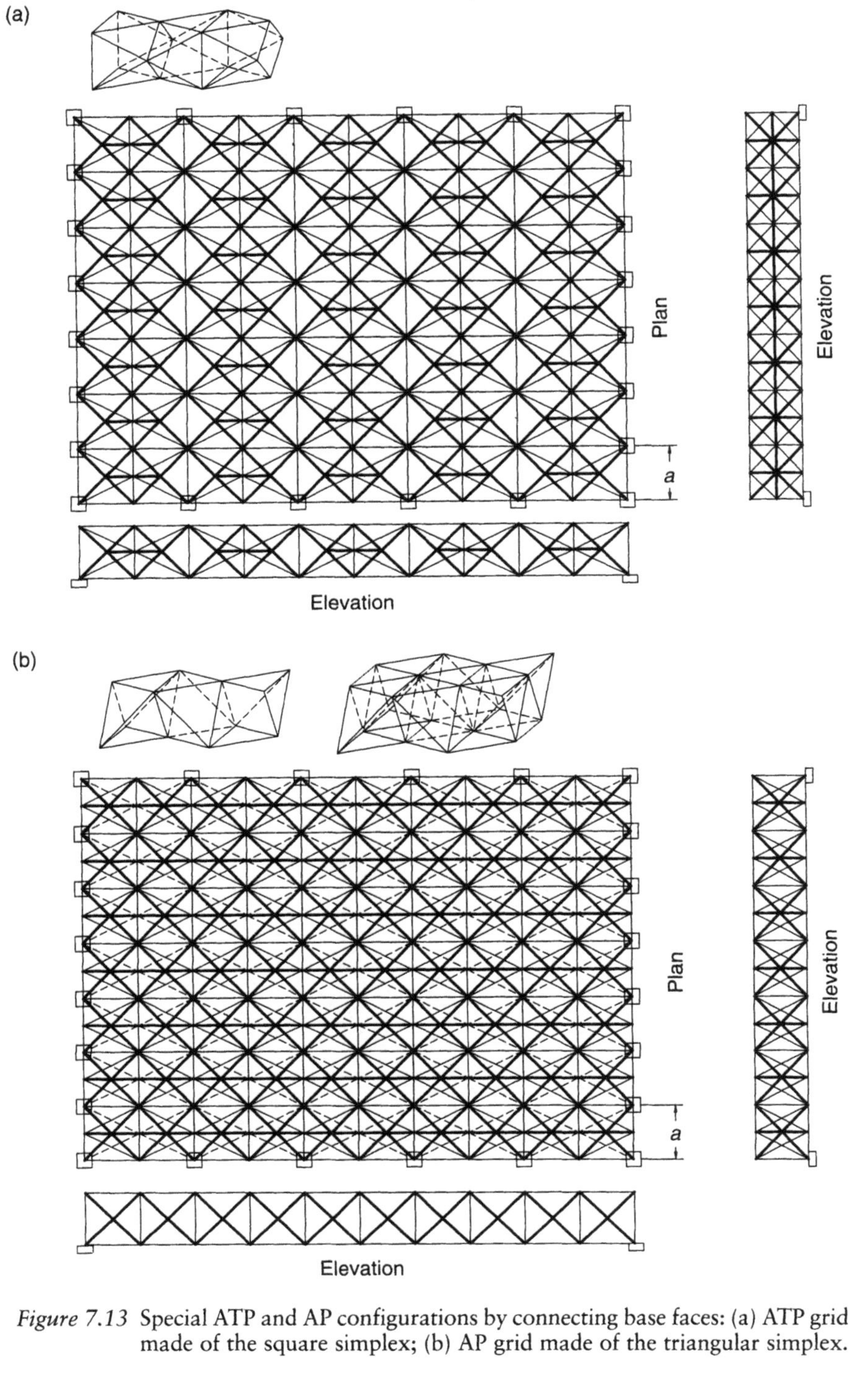

Figure 7.13 Special ATP and AP configurations by connecting base faces: (a) ATP grid made of the square simplex; (b) AP grid made of the triangular simplex.

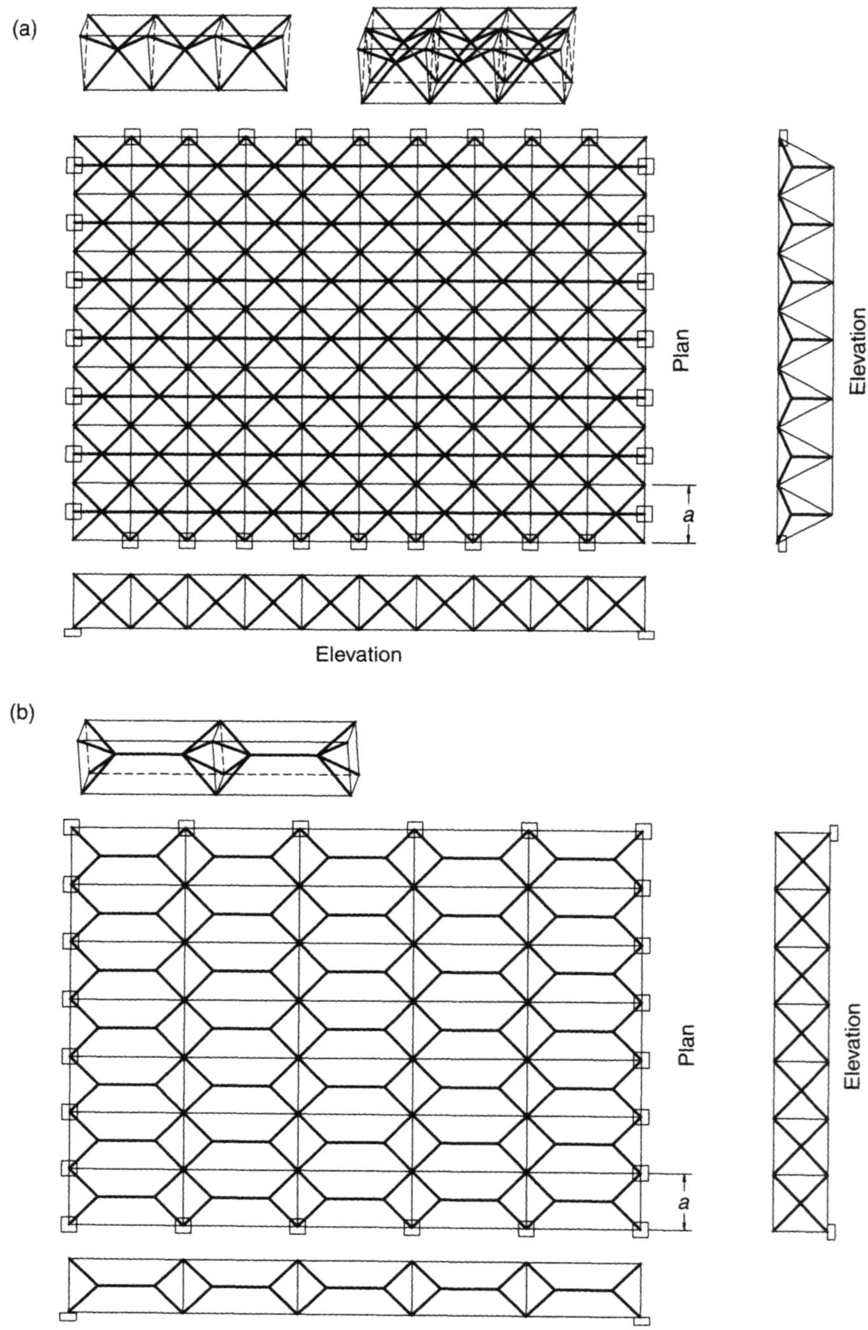

Figure 7.14 Special P and P2 configurations by connecting base faces: (a) P grid; (b) P2 grid.

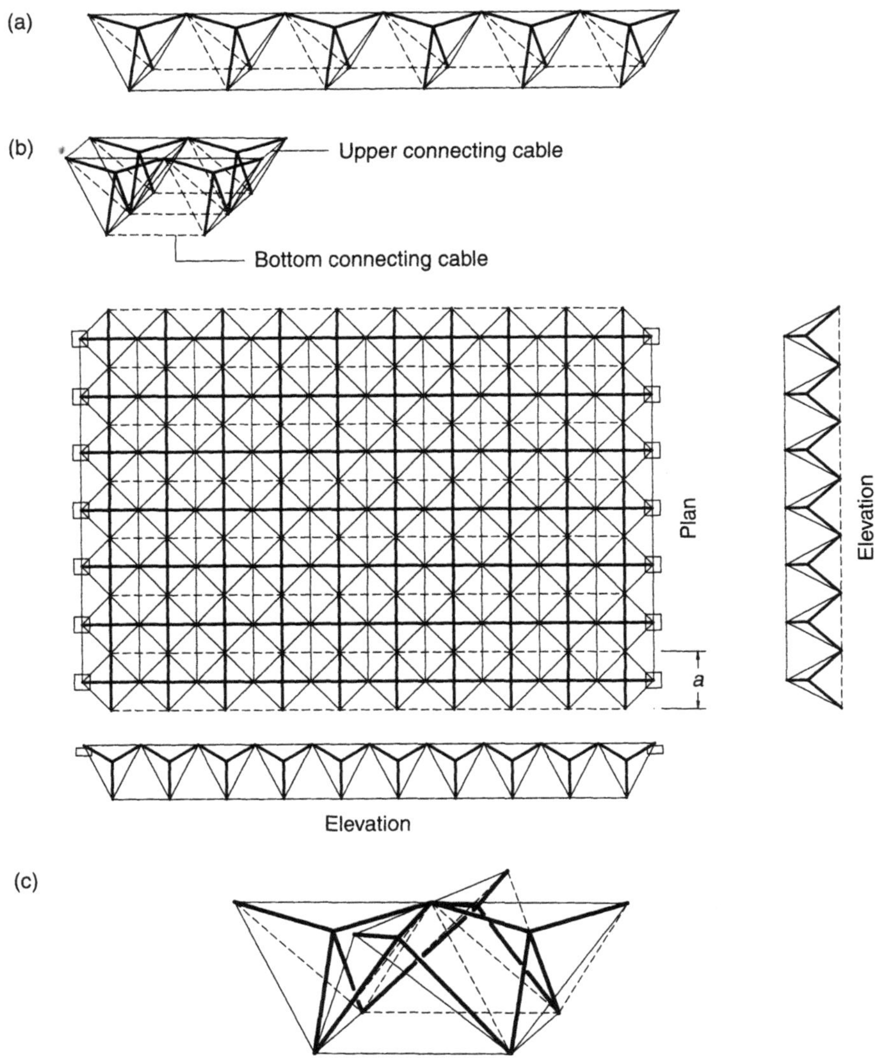

Figure 7.15 Special CP configurations by connecting edges: (a) a CP truss; (b) a CP grid; (c) two-way configuration of (a).

7.4.1 Hyperbolic paraboloid membrane roof

Application of hyperbolic paraboloid membrane in polyhedra has been studied (Hooper 1998). Here, the membrane can be attached easily to a CP. In Figure 7.16, it is attached to four edges of a four-vertex CP. A small roof as shown is composed of four CPs, whose array is presented in Figure 7.15(b). More arrays of CPs are feasible.

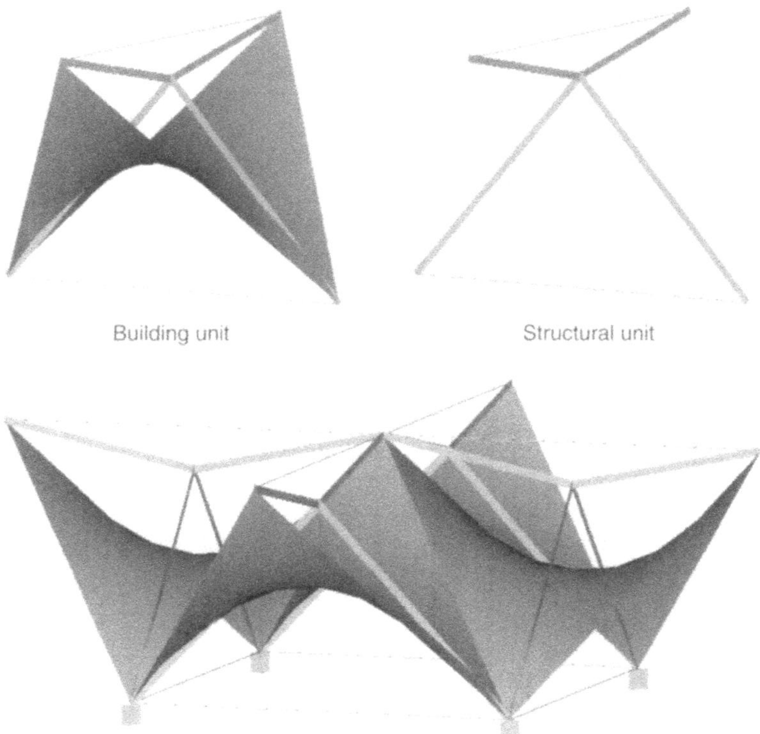

Figure 7.16 Hyperbolic paraboloid membrane covering CPs.

7.4.2 *Two tents*

As a 'special' way of applying cable-strut simplexes, a five-vertex CP supporting a membrane tent is supported by three vertices of a triangular face at the ground (Figure 7.17(a)). Two free vertices extend out the volume below covered by the membrane. A one-way array of tents can be connected successively to cover wider area. The concept shows that a simplex can be supported in different ways when necessary.

As another type of tents, each structural unit is composed of two connected quadrilateral-based DPs (Figure 7.17(b)). Each pair of DPs is supported to the ground at the bottom joint, forming truss-like mechanism. Discussion of structural detail is omitted for simplicity. Note that in addition to the membrane tent, other shapes can be attached to the skeleton, including the 'sail-like' form (Berger 1997).

7.4.3 *A petal roof*

The concept of 'petal shell' may not be new to architects (Pavlov 2002). The invention of cable-strut systems provides more choice of design. For example,

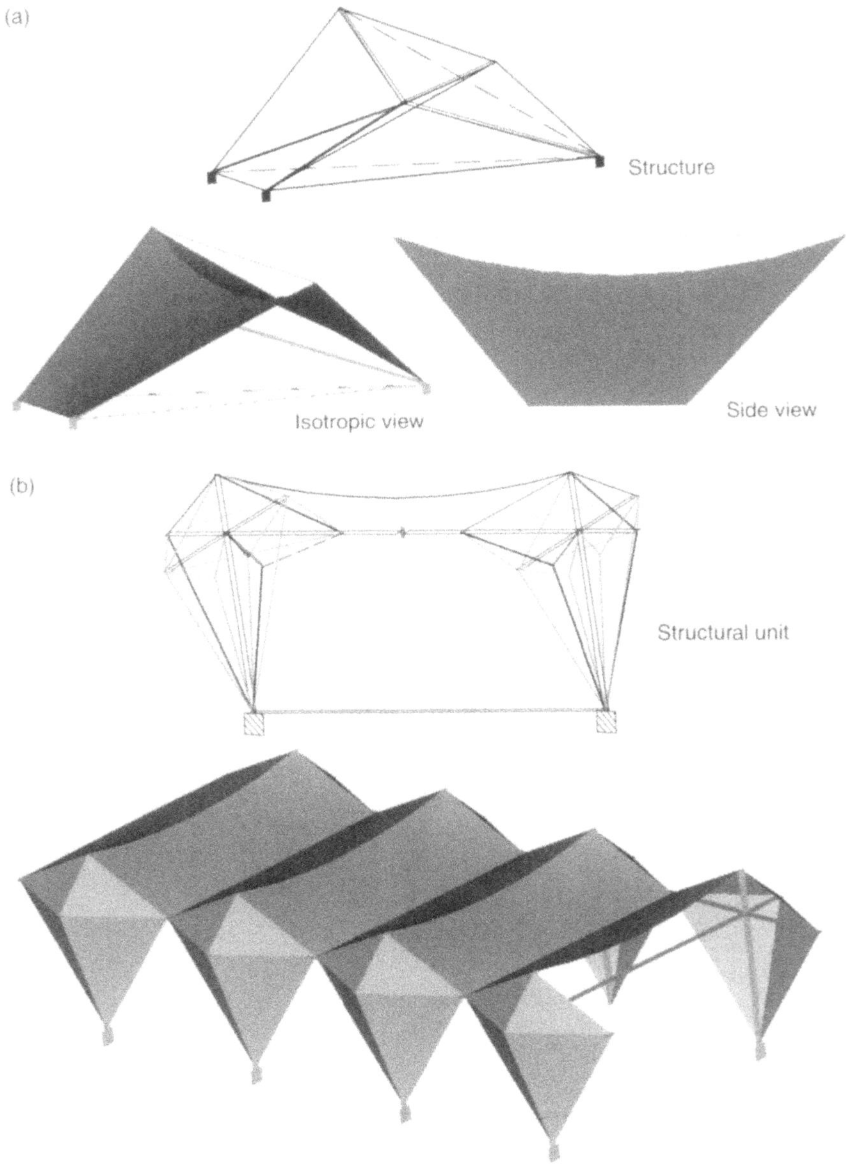

Figure 7.17 Tents made of cable-strut simplexes: (a) a tent of CP skeleton; (b) a tent of DP skeleton.

flexible membrane petal can be used to enclose the volume and the membrane is attached to an octagonal AP (anti-prism), as shown in Figure 7.18. In order to enclose bigger volume, lower inclined struts can be curved, and in addition, diagonal cables can be curved by tying them to upper cables.

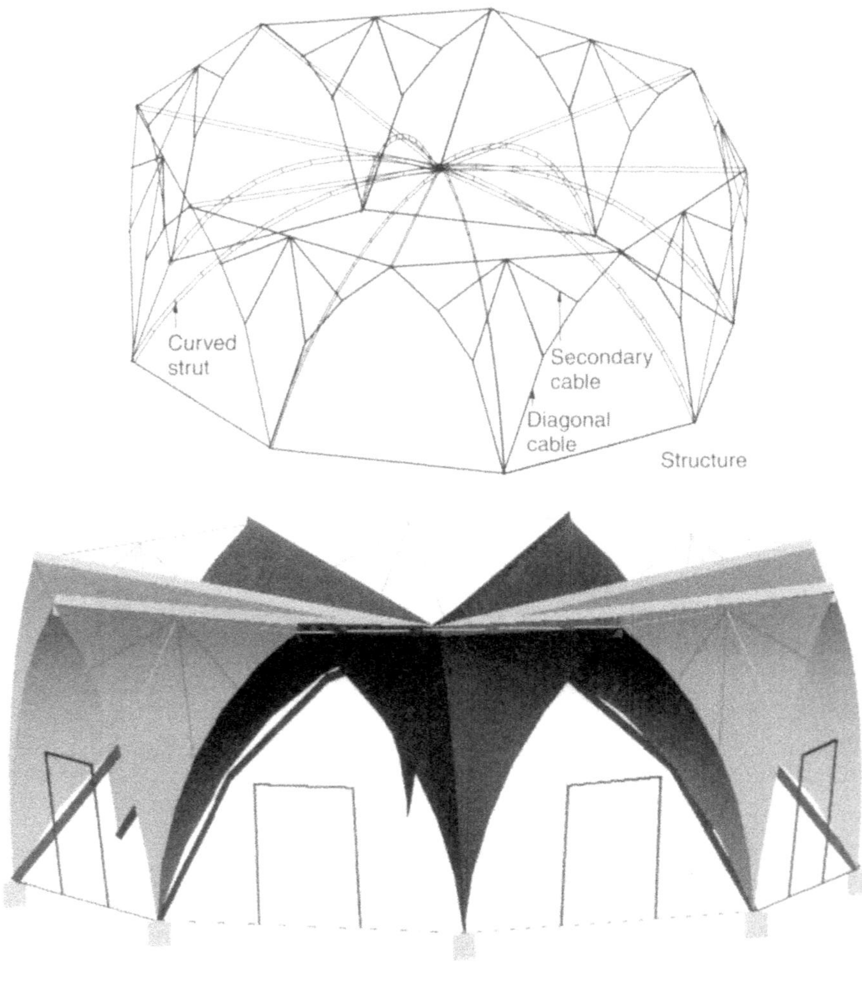

Figure 7.18 A 'petal shell' of AP skeleton.

7.5 Basic cable-strut structures with ornamental supplements

As the simplest architectural design, ornamental supplements are added to cable-strut simplexes. The supplement can be either special-shaped panels or components like tendons and bars without any effect on structural behaviour.

7.5.1 Basic simplexes with ornamental panels

Based on the concept, ornamental panels are attached to cable-strut simplexes to present aesthetics. In Figure 7.19(a), it is presented as an array

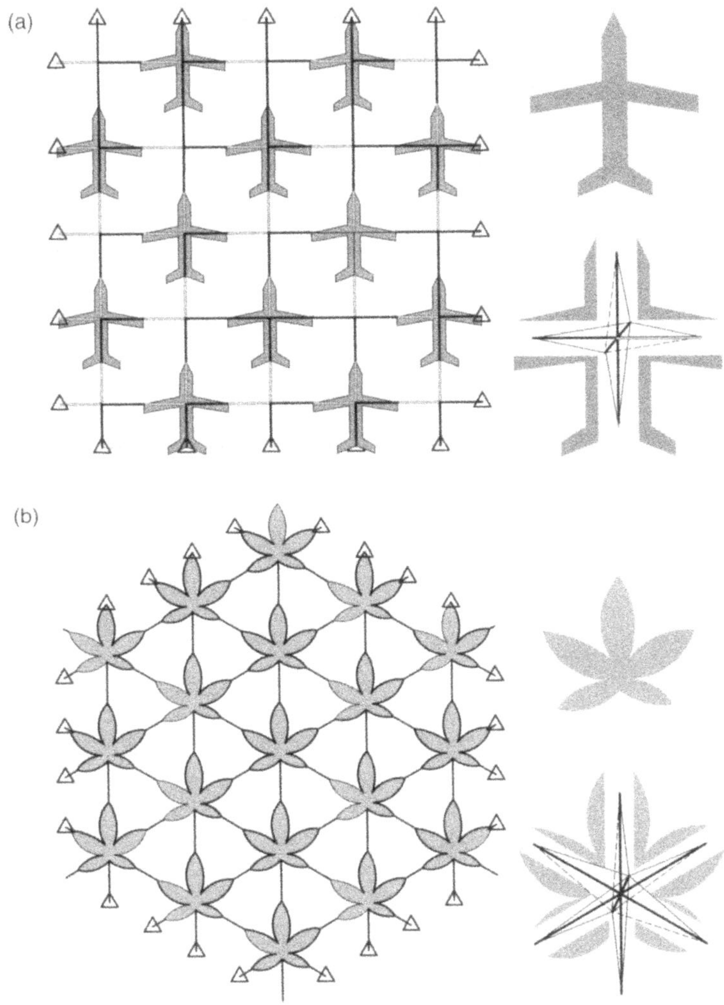

Figure 7.19 DP grids with ornamental panels: (a) square DP with airplane; (b) hexagonal DP with maple leaf.

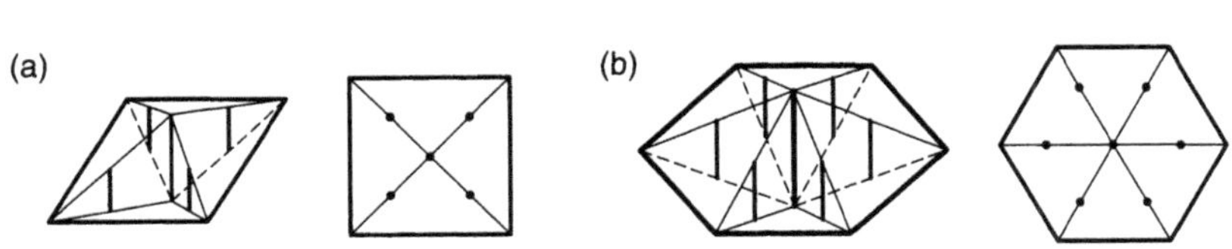

Figure 7.20 RP simplexes with supplementary bars: (a) square RP; (b) hexagonal RP.

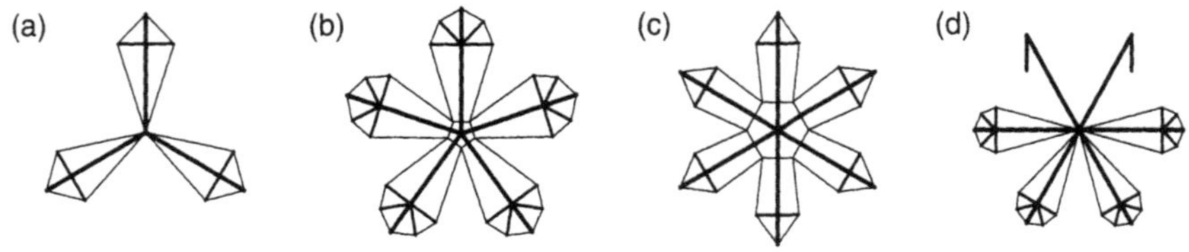

Figure 7.21 DP simplexes with supplementary components in the base: (a) a DP propeller; (b) a DP flower; (c) a DP snowflake; (d) a DP butterfly.

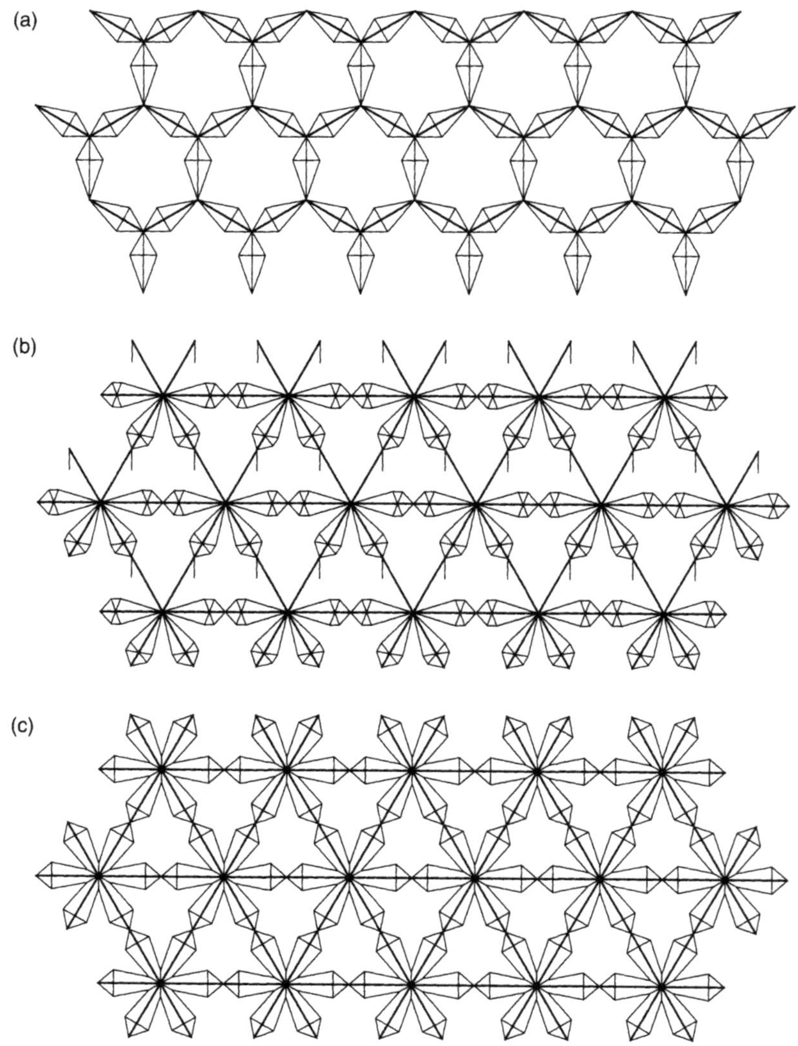

Figure 7.22 Array of DP simplexes with supplementary components: (a) an array of DP propeller; (b) an array of DP butterflies; (c) an array of DP snowflakes.

of airplane models attached in the strut layer of DP grids. Each 'plane' is composed of four pieces of elements, connected to a square-based DP. In Figure 7.19(b), it is illustrated as an array of five-panel maple leaf, attached to the strut layer of DP grids composed of hexagonal simplexes. Shapes as many as one may wish, such as a butterfly, an insect, a fish and a flower, etc. can be introduced.

Extensively, ornamental panels are applicable to the strut layer of RP grids, CP grids, the lower inner struts of ATP (anti-truncated-pyramids), AP and P grids, or to the bottom cable layer of any cable-strut grids, presenting different eyesight effect.

7.5.2 *Basic simplexes with small additional components*

Supplemental components can be added to basic cable-strut simplexes to form funny units of different appearance. As the example for an RP, additional bars can be attached to inclined cables (Figure 7.20), and the resulting shape resembles radial cable roof.

In DP simplexes (or CP simplexes), a triangular becomes a propeller (Figure 7.21(a)), a pentagonal becomes a flower (Figure 7.21(b)), and a hexagonal becomes a snowflake or a butterfly (Figure 7.21(c), and (d)), in which additional short bars and cables are attached to the struts in the base. These derived forms present special appearance in grids (Figure 7.22), especially a butterfly (Figure 7.23) built from a number of 'flowers'.

7.6 Basic cable-strut structures with roof sculptures

Based on the concept, structural art is expressed through sculptures of specially shaped roof. It may be considered as the extended concept from Section 7.4. Modified design of supporting simplexes or grids may be required to match roof design. Note that the concept is feasible benefiting from the cables in simplexes, which are convenient for attaching and prestressing.

7.6.1 *Application in simplex structures*

Discussion of the concept starts from simplex structures in this section. A simple example is a maple leaf attached to an octagonal CP (Figure 7.24). The leaf part is covered by red membrane, whereas the boundary non-leaf part is covered by white coloured membrane or even glass. The leaf can be seen from both the inside and outside of the enclosure. In order to follow the maple shape, the vertical strut is moved away from the centre.

A more complex example is shown in Figure 7.25, in which the transformation from the basic simplex is required in order to meet the architectural form. In Figure 7.25(a) an octagonal TP2 (truncated pyramid with two inner joints) is presented, in which lateral cables and supports in the base

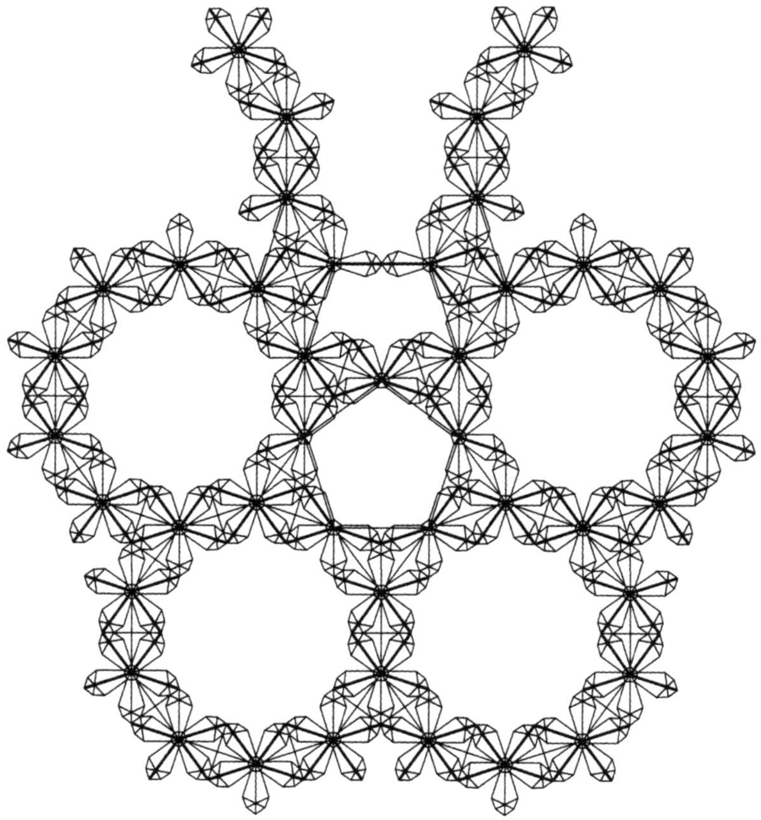

Figure 7.23 A flower butterfly made of DP flowers.

remove mechanisms and provide sufficient stiffness. In Figure 7.25(b), the positions of eight top joints are adjusted so that seven joints are located along a semi-circle. The inclined strut connected to the left joint is modified to a broken-line shape (taking moment as an occasional case) and the vertical strut is moved away from the centre. The transformed simplex allows a membrane peacock attached to the upper inclined strut skeleton as the sculpture and the membrane depicted with a maple leaf attached at the lower inclined strut skeleton as the structural roof for the peacock to stand on (Figure 7.25(c)).

The detail of the form is illustrated in Figure 7.26. In the upper sculpture part, the head and body of the peacock is made of hydrogen-inflated membrane, tied to the broken-line strut. The feather tail is portrayed in each piece of white membrane attached to inclined struts and upper base cables. Under wind flow, the peacock looks alive following the vibration of the

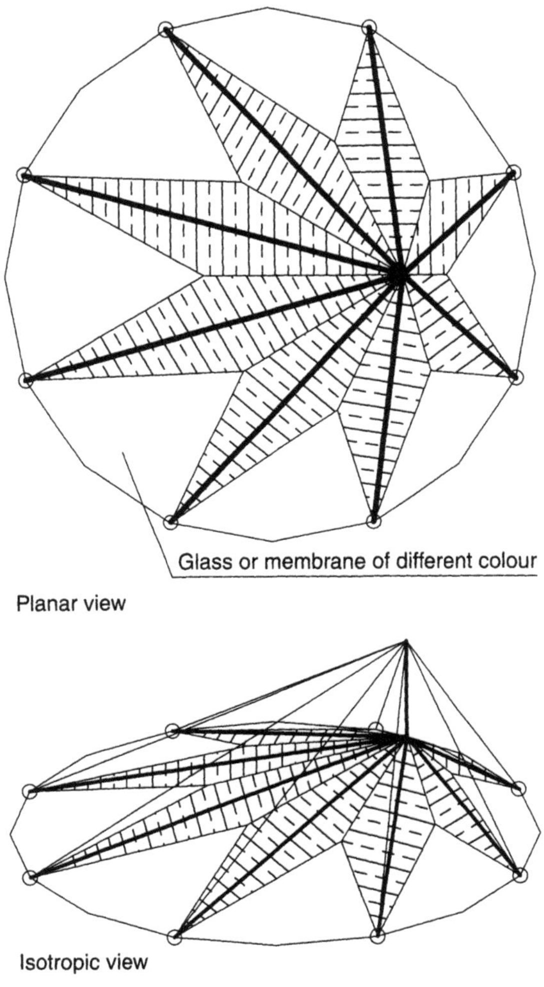

Figure 7.24 A maple leaf made of an octagonal star CP.

membrane body! The lower structural part attached with maple leaf membrane is not difficult to cover the 30m span.

As the further application, the one simplex form can be connected to form twin-simplex or multi-simplex form. In Figure 7.27, it is shown that further application of a triple-simplex form. The sculpture part includes two bigger peacocks facing each other standing on the maple leaf roof and a small orthogonal peacock standing on a small maple canopy. The door is below the half-cantilevered small maple leaf and the enclosure along the boundary is indicated in the figure. The form covers an elliptical layout sufficient for a small stadium.

Figure 7.25 A new architectural form incorporating roof and structural skeleton: (a) an octagonal TP2; (b) specially designed skeleton; (c) supported membrane roof of maple leaf and decorated peacock – a peacock standing on a maple leaf.

Source: Wang and Li (2003b).

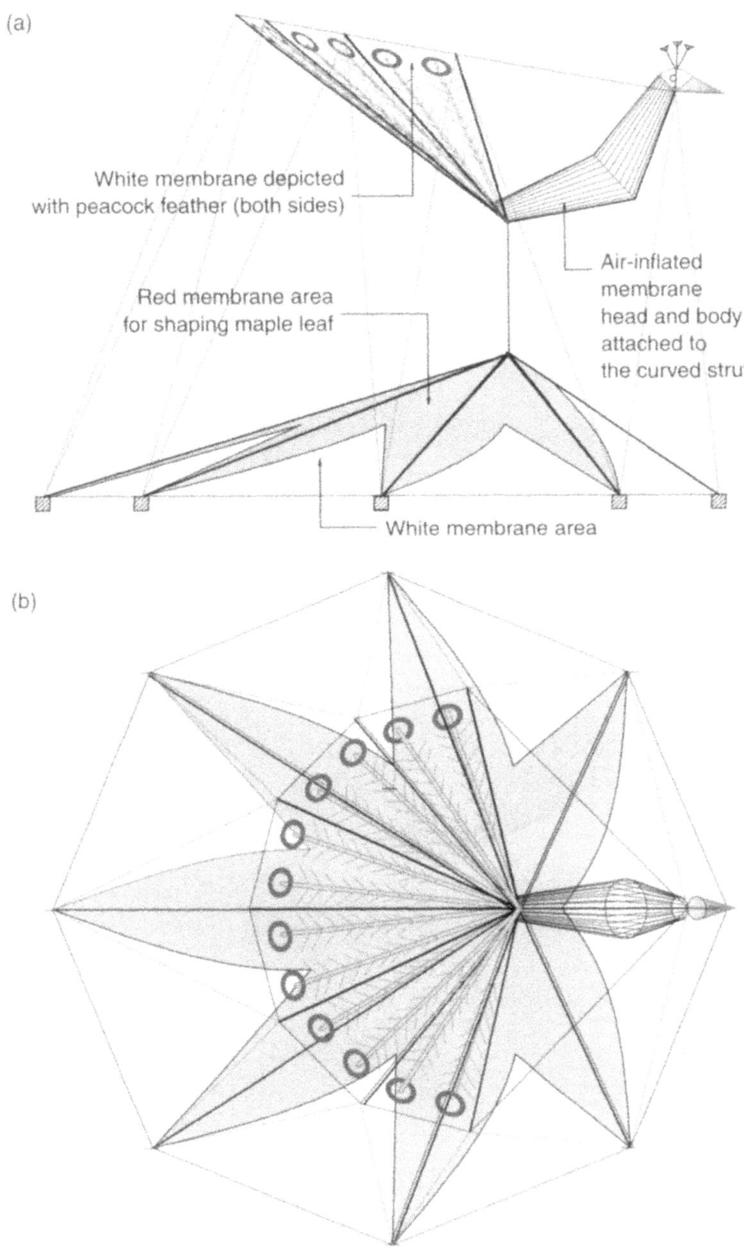

Figure 7.26 Layout of the design in Figure 7.25: (a) elevation; (b) plan.
Source: Wang and Li (2003b).

(a)

(b)

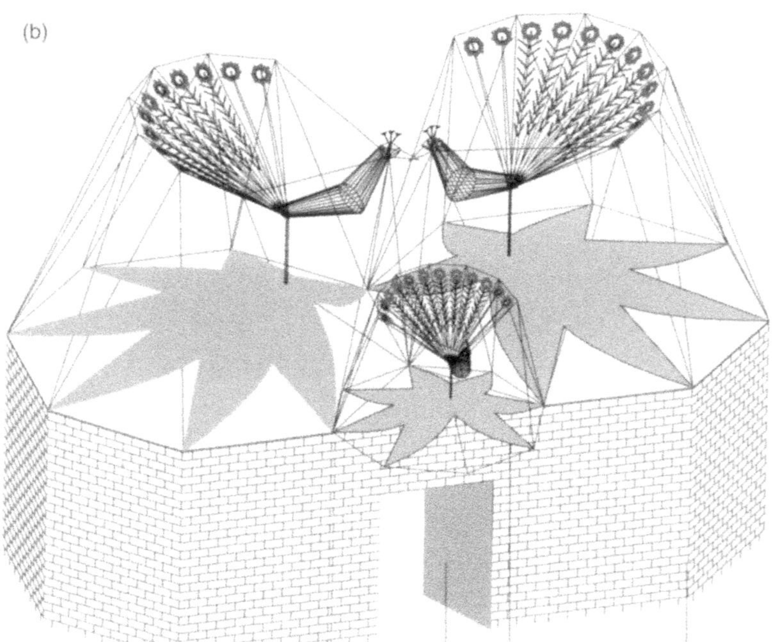

Figure 7.27 Triple simplex form developed from Figure 7.25: (a) front view; (b) layout.

7.6.2 Application in grids

In grid application, a wide variety of roof sculpture forms are applicable, including an airplane, a bird, an insect, and a flower, etc. The 'actual' roof may have to be attached to the bottom layer of the grid. However, it shall be borne in mind that the shape of the sculpture should match the structural shape of the simplex as the design technique and how to 'attach' it should be considered.

In Figure 7.28, a flying swan is connected to the upper four inclined struts of the square ATP following the shape of two wings and body. Two additional examples, presented as the expression of aesthetics and various ways of connecting roof sculpture, are birds resting on a DP grid (Figure 7.29) and dragonflies resting on a RP grid (Figure 7.30). In the former case, each bird is attached to three horizontal struts and the upper vertical strut of the square DP. In the latter, a dragonfly has to rest respectively its body on two horizontal struts and its wings on four upper inclined cables from two adjacent RPs.

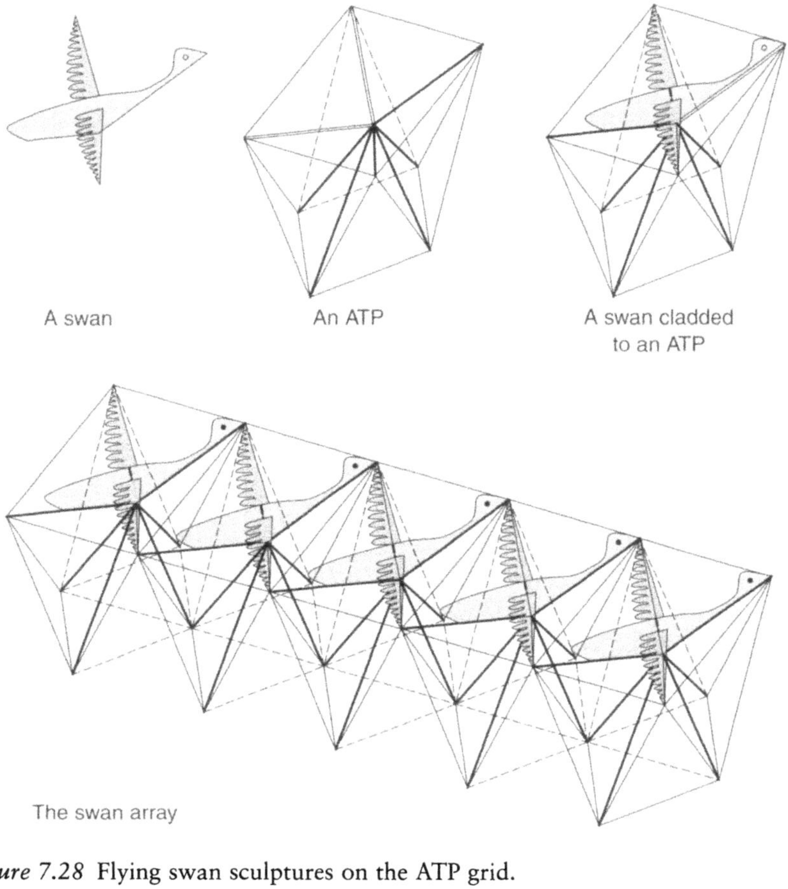

Figure 7.28 Flying swan sculptures on the ATP grid.

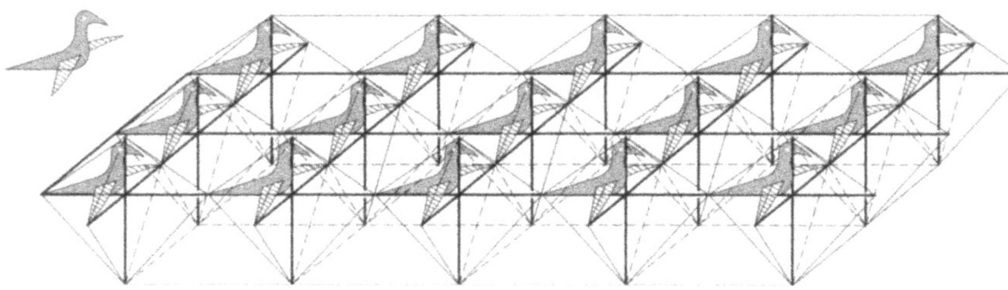

Figure 7.29 Small birds resting on the struts of the DP grid.

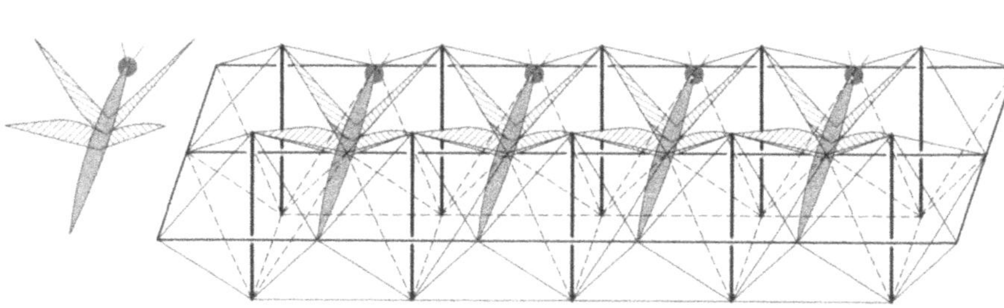

Figure 7.30 Dragonflies resting on the horizontal struts and inclined cables of the RP grid.

7.7 Reshaped cable-strut simplexes

Based on the concept, aesthetics is demonstrated from structure itself through designing special geometry to simplexes. The reshaping can be applied to base cables, edge cables or struts. Some examples, among possible choices, are presented as follows.

7.7.1 Star simplexes family

Stars in one plane

Cable-strut simplexes can be shaped into their corresponding star forms when their bases are star-reshaped. In the star-shaped simplexes built from a CP, a DP or a P (TP), star angles are shaped by broken-lined base cables stabilized by tying additional cables from inside. The tying cables may be enclosed among themselves (Type I) or connected to a common node in the plane (Type II), presenting different appearance. The resulting simplex may be named a star-shaped CP (SCP), a star-shaped DP (SDP) and a star-shaped P (SP), etc.

A CP contains a state of self-stress and no mechanisms. In a triangular SCP, Type I (Figure 7.31(a)), three additional infinitesimal mechanisms acting normal to the base plane are produced. One mechanism occurs at each

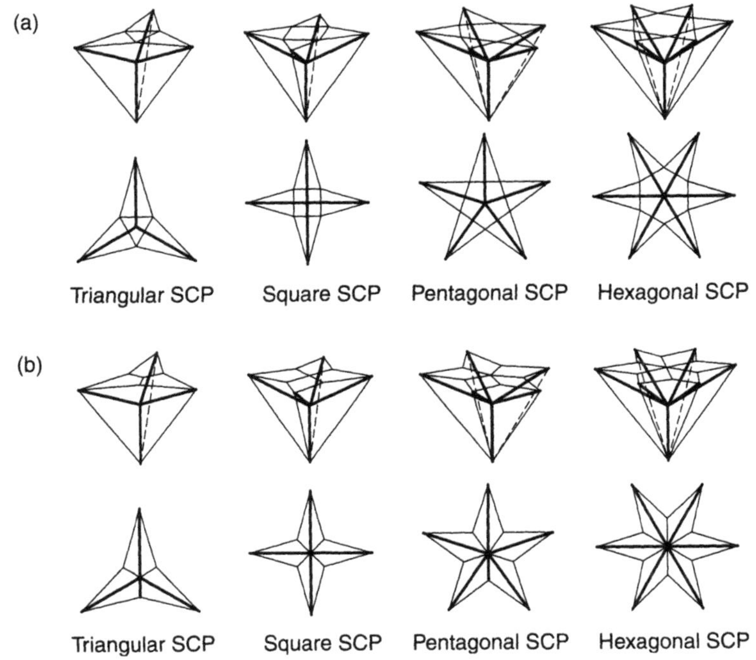

Figure 7.31 Star CP simplexes: (a) Type I; (b) Type II.

vertex of the small triangle that is enclosed by additional cables in the base. In a CP higher than the triangular, an infinitesimal mechanism of in-plane rotation is produced with a state of self-stress in the small polygon. All these first-order mechanisms can be stabilized by self-stress. Actually, these mechanisms can be ignored, as they are unlikely to be activated by actual load to affect structural behaviour. When a CP is transformed into an SCP, Type II (Figure 7.31(b)), the common joint connecting tying cables just adds three ignorable infinitesimal mechanisms of translation, compared with Type I. So the properties of two types of SCPs are basically the same and are similar to those of the corresponding CP.

P simplexes can be reshaped into SPs or SP2s, but the discussion is focused on SP for simplicity. A P contains a self-stress state, an infinitesimal mechanism of prism rotation, and n-3 state of finite mechanisms of in-plane distortion in the bases (e.g. one for the square), where n is the number of edges in the base. When it is transformed into an SP, Type I (Figure 7.32(a)), $2n$ new infinitesimal mechanisms acting normal to the base plane are produced, n in each base. Meanwhile, one new infinitesimal mechanism of in-plane rotation acting within each base plane (altogether two) is produced accompanying a new self-stress state in each base (altogether two) for all SPs. As all increased infinitesimal mechanisms can be ignored, the properties of an SP, Type I are similar to those of the corresponding P. The case

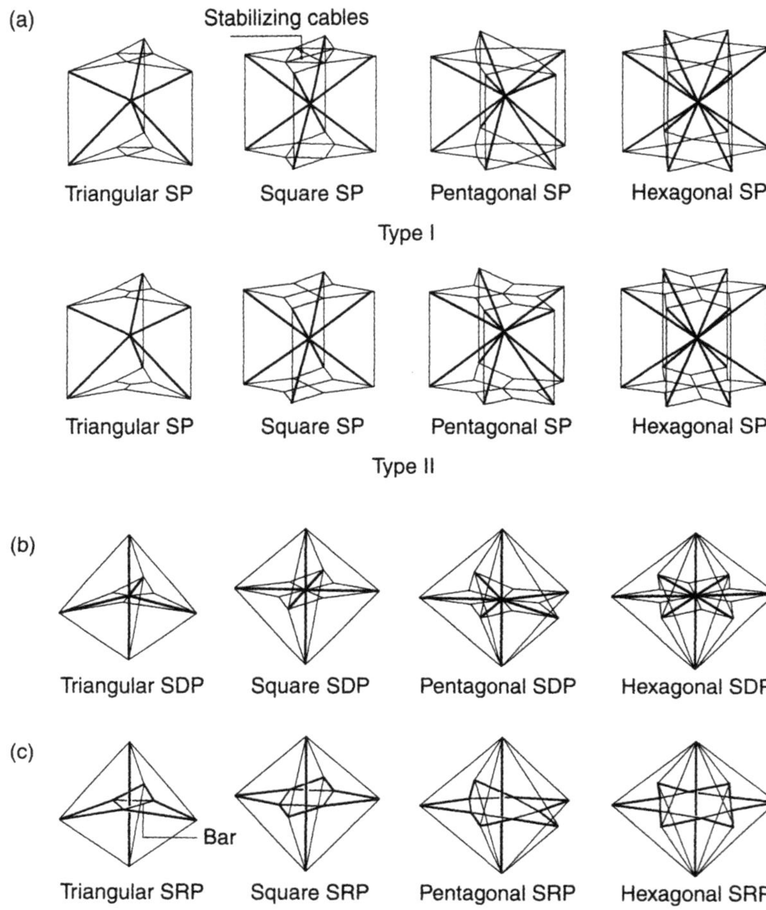

Figure 7.32 Other star cable-strut simplexes: (a) star P simplexes; (b) star DP simplexes (Type II); (c) star RP simplexes.

is the same for an SP, Type II (Figure 7.32(a)). Like the case for a P when $n > 3$, additional cables are also required to stabilize the finite mechanisms. For example, two additional crossing cables are added in the base of the square SP (Figure 7.32(a)).

An AP or ATP is star form itself. To make sharper star angles, the same design as in P simplexes can be applied to their bases.

An SDP, Type II (Figure 7.32(b)) contains only one layer of tying cables, which are connected to the inner joint directly. Hence, only n infinitesimal mechanisms normal to the base plane are produced.

A star-shaped RP (SRP, Figure 7.32(c)) is 'special' compared with other simplexes. The star angles are formed by broken-lined struts in the base.

Meanwhile, these struts shall be connected by bars for stability. As the strut layer is compressive, joints connecting stabilizing bars should be designed for moment resistant for out-of-plane stability.

The grid configurations of star simplexes can refer to those of the corresponding basic simplexes, although all the resulting forms look like edge-to-edge types (Figure 7.33(a)), presenting aesthetics through structure themselves. Note that like their basic simplexes, star forms also have linear, plane-filling, space-filling forms, and special configurations in Section 7.3, etc. A plane-filling form made of pentagonal star simplexes in Figure 7.33(b) and a special configuration made of the hexagonal SDP in Figure 7.33(c)

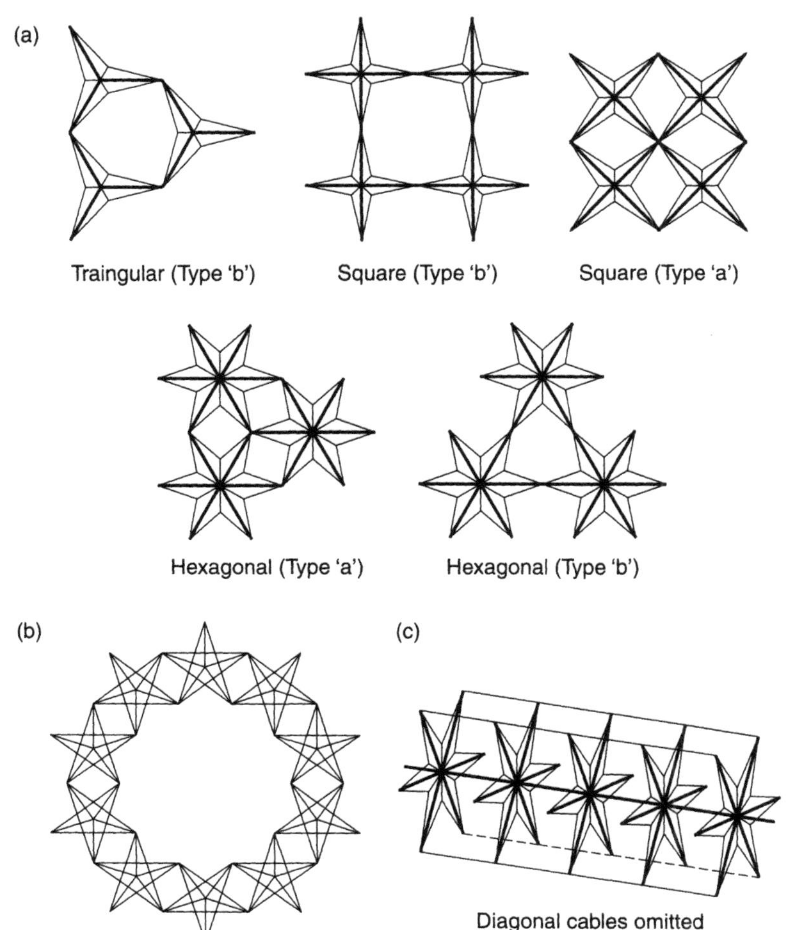

Figure 7.33 Configurations made of star cable-strut simplexes: (a) basic configurations; (b) plane-filling form made of pentagonal simplexes; (c) special configuration made of the hexagonal SDP.

are given as examples. In addition, special roof or sculpture design can also be introduced to present more exciting forms.

Stars in all ways

In this discussion, stars are introduced in the base plane as 'planar' stars. The concept can be extended to all vertices to form 'spatial' stars. Therefore, every edge cable is tied inside by the introduction of tying cable, as illustrated in Figure 7.34. The resulting forms are named by adding

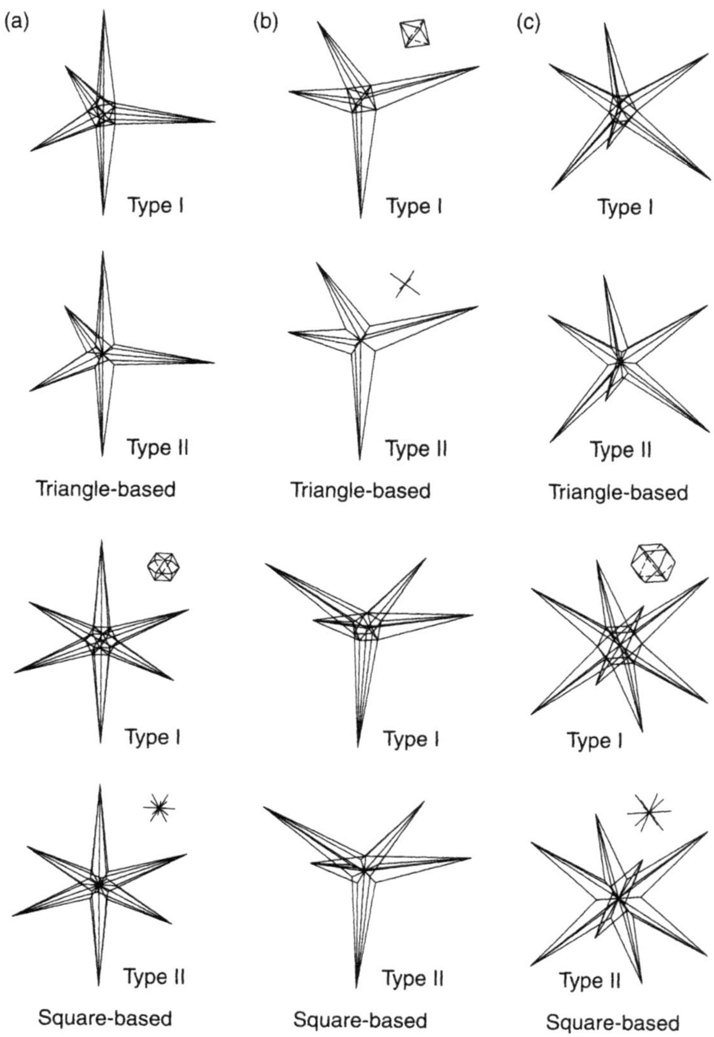

Figure 7.34 Star cable-strut simplexes in 3D form: (a) SDP-3D; (b) SCP-3D; (c) SP-3D.

suffix '3D'. It seems that lower simplexes (those containing less base cables) are more suitable for the concept.

Similar to the case in 'planar' forms, tying cables in spatial star forms may be enclosed among themselves (Type I) or connected to the inner joints of the simplex (Type II). In Type II, each tying cable lies in the plane shared by the broken-lined edge cable and two coplanar struts. Examples of triangle- and square-based SDP-3D, SCP-3D and SP-3D are illustrated in Figure 7.34(a), (b) and (c), respectively. Note that in Type I, tying cables enclose small polyhedra, including a cub-octahedron in the square-based SDP-3D (Figure 7.34(a)) and SP-3D (Figure 7.34(c)), an octahedron in the triangle-based SCP-3D (Figure 7.34(b)), etc.

All configurations applicable for planar star simplexes are naturally suitable for spatial star simplexes. 'Unique' looking can be presented from these simplexes as exemplified from the planar/spatial array of square-based SDP-3D in Figure 7.35, and the SDP column made of the radial array of pentagon-based SDP-3Ds in Figure 7.36, etc.

7.7.2 Curved strut forms

In the section, the concept is illustrated through designing curved struts. Two types of curved strut forms are proposed here. One is the 'inner circle' type that the circle is shaped inside the simplex. When the concept is applied to a DP, the struts in the base are connected to an additional strut ring, in place of connecting directly to the vertical struts (Figure 7.37). Therefore, the original two vertical struts become one member. New appearance is presented at the expense of reduced structural efficiency. When the concept is applied to a P (Figure 7.38), an ATP or an AP, inclined struts are connected to the central strut ring in place of a common joint.

The other is the 'outer circle' type that the circle is shaped outside the simplex. It is more suitable for RP simplexes. As shown in Figure 7.39, struts in the base are reshaped to an arc. Arc-shaped struts from adjacent simplexes enclose a circle in the grid. Six adjacent simplexes are required for the triangular, presenting largest circle (Figure 7.39(a)); four for the square, presenting smaller circle (Figure 7.39(b)); and three for the hexagonal, presenting smallest circle.

7.7.3 'Bicycle-wheel' and 'cable-dome' types

An RP can be reshaped by replacing the vertical strut to a circular array of struts, presenting 'bicycle-wheel' shape (Figure 7.40). The concept is grafted from a bicycle-wheel cable roof. The square-based bicycle wheel (Figure 7.40(a)) contains a state of self-stress and seven mechanisms. Among the seven, two infinitesimal mechanisms are rotation of two cable rings, other three being in-plane distortion of the vertical square enclosed by two vertical struts and two horizontal cables (the left in-plane distortion is the linear combination of the other three). The left two finite mechanisms

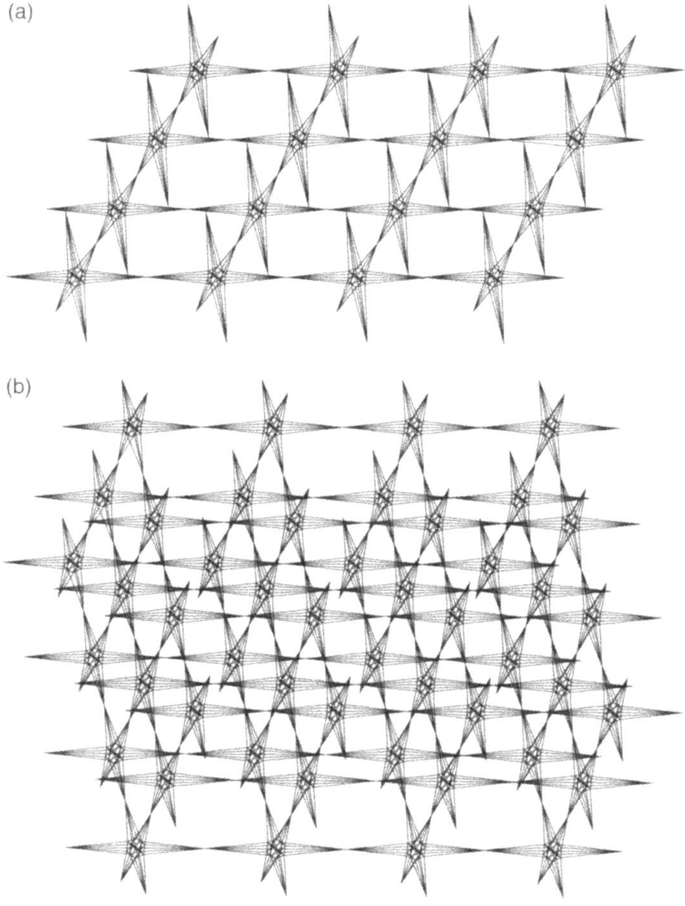

Figure 7.35 Array of SDP-3Ds: (a) planar array; (b) spatial array.

are rotations in the outer square strut ring, hence, the strut ring should be designed moment resistant for stability (despite that the moment is negligible). The properties of a hexagonal type are similar. The resulting grids are given for reference (Figure 7.40).

Alternatively, an additional vertical strut with diagonal cables is attached to the upper cable ring (Figure 7.40), resulting in 'cable-dome' shape with pyramidal faceted surface shown on the roof. The way of forming grids is the same as 'bicycle-wheel' type.

7.8 Higher cable-strut polyhedra

Higher cable-strut polyhedra are unlikely to be as efficient as building blocks for structurally efficient grids. But they present unique appearance in

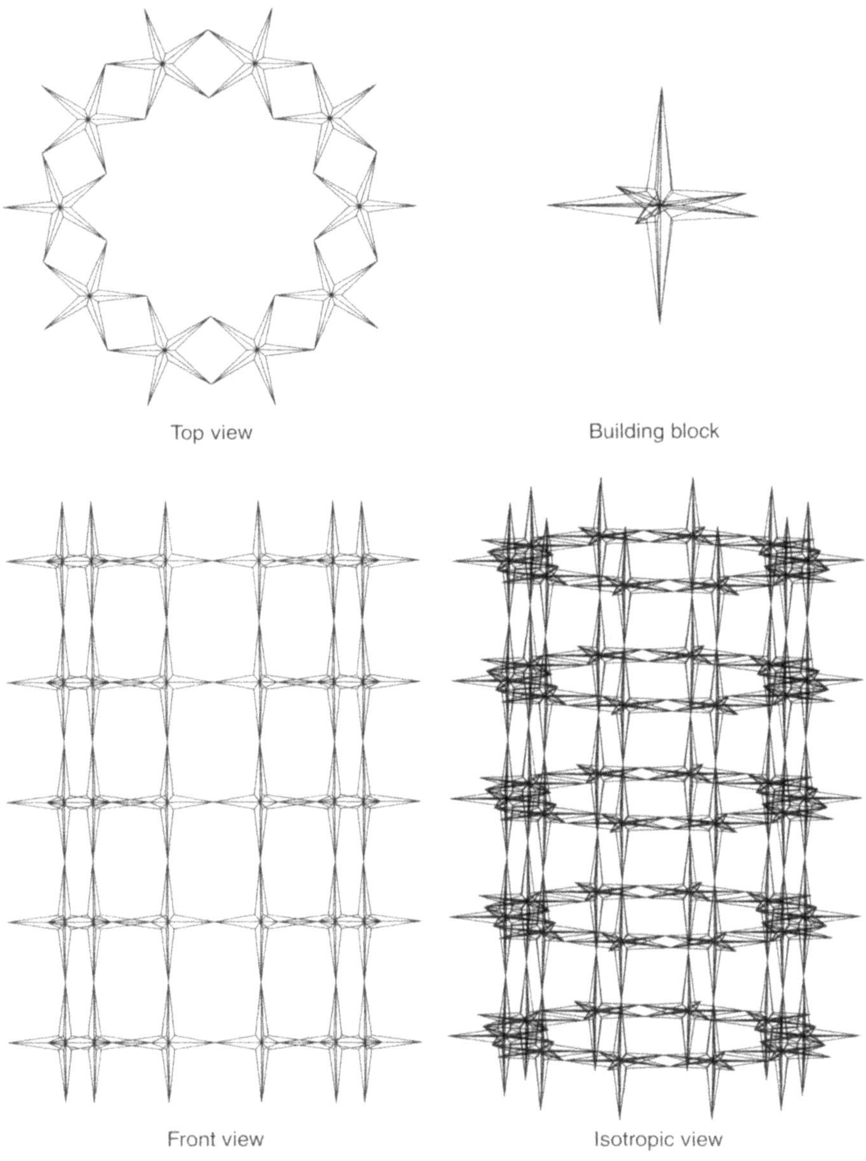

Figure 7.36 A star column made of pentagonal SDPs.

architecture. In order to form polyhedra of cable-strut types, the geometry is normally stabilized by struts from inside and cables as edges. But it is also possible to design struts in the outer layer in some relatively simple polyhedra. One such case is the derivature of RP with two strut rings forming the base of the derived prisms, as shown in Figure 7.41(b).

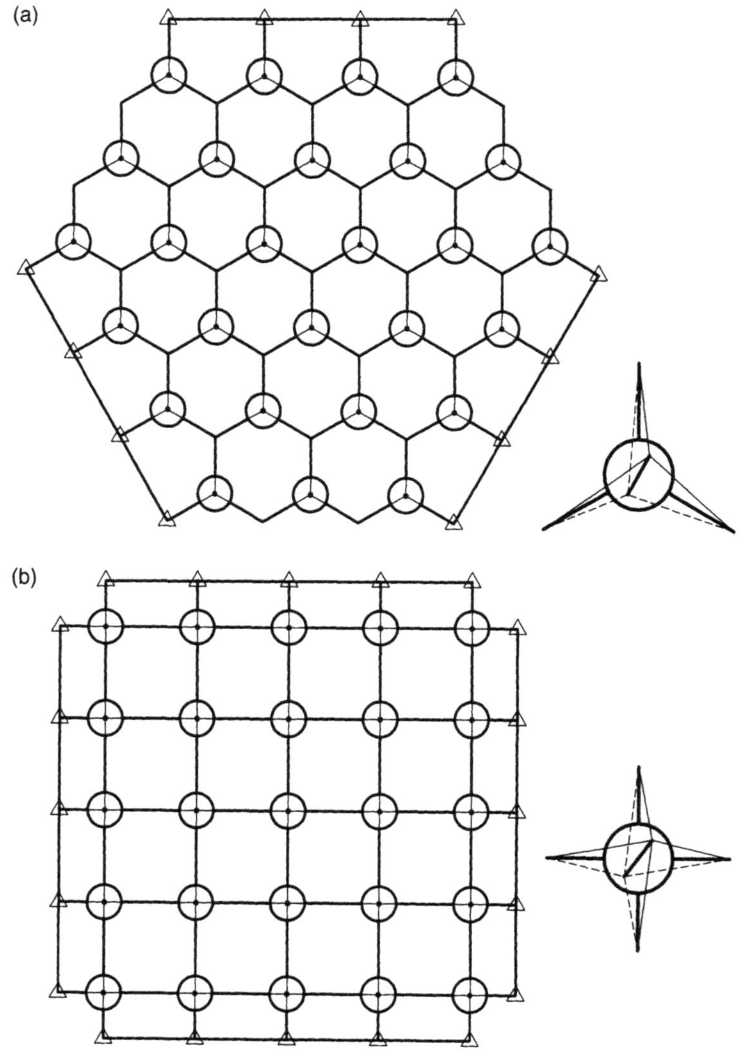

Figure 7.37 Curved strut DP form – 'inner' circle type: (a) triangular DP type; (b) square DP type.

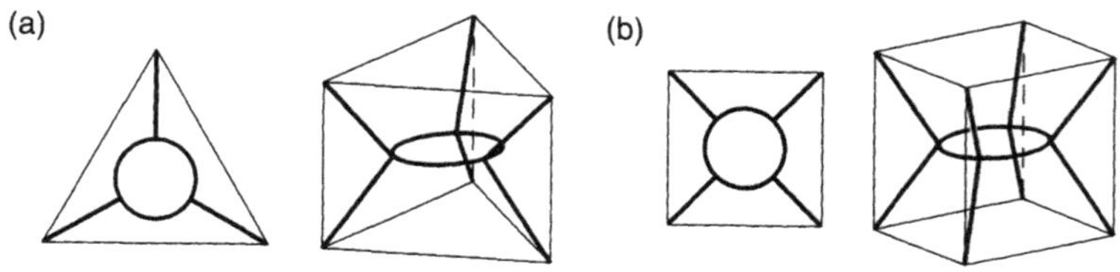

Figure 7.38 Curved strut P simplexes: (a) triangular P; (b) square P.

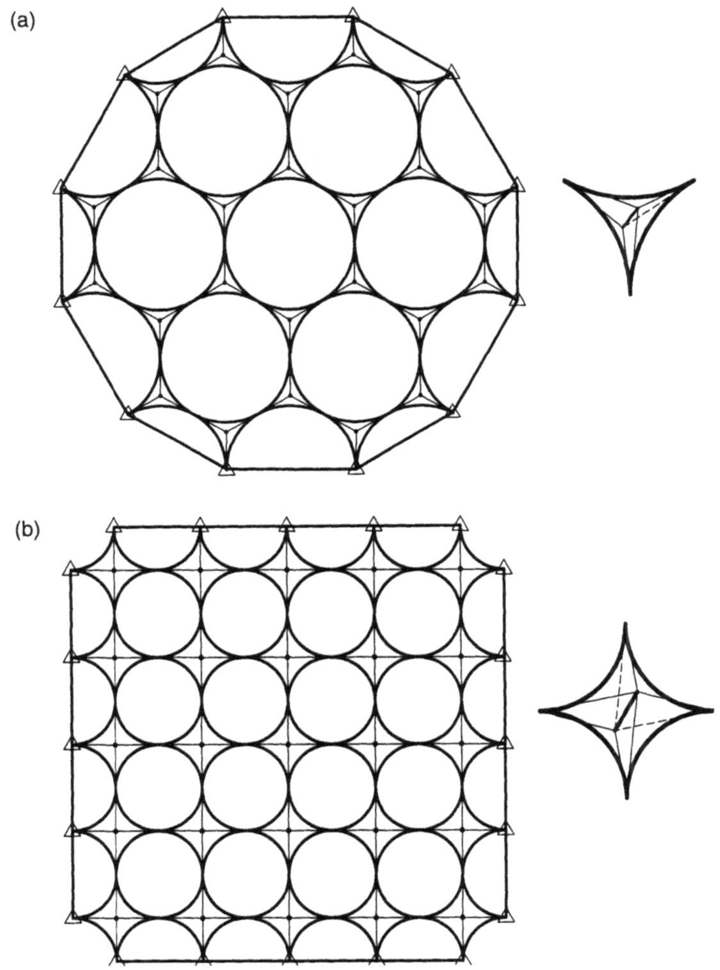

Figure 7.39 Curved strut RP form – 'outer' circle type: (a) triangular RP type; (b) square RP type.

When no struts are designed in the outer layer, struts may join at the centre directly provided that the polyhedra is not quite high. For example, the geometry in Figure 7.41(a), which is hence considered a DP derivature, can be stabilized by designing struts joined at either one or two inner joints (Figure 7.41(c) and (d)).

In order to achieve short struts especially in high polyhedra, the general principles are presented as follows:

- One vertex is connected to only one strut. It means each vertex is only used once.

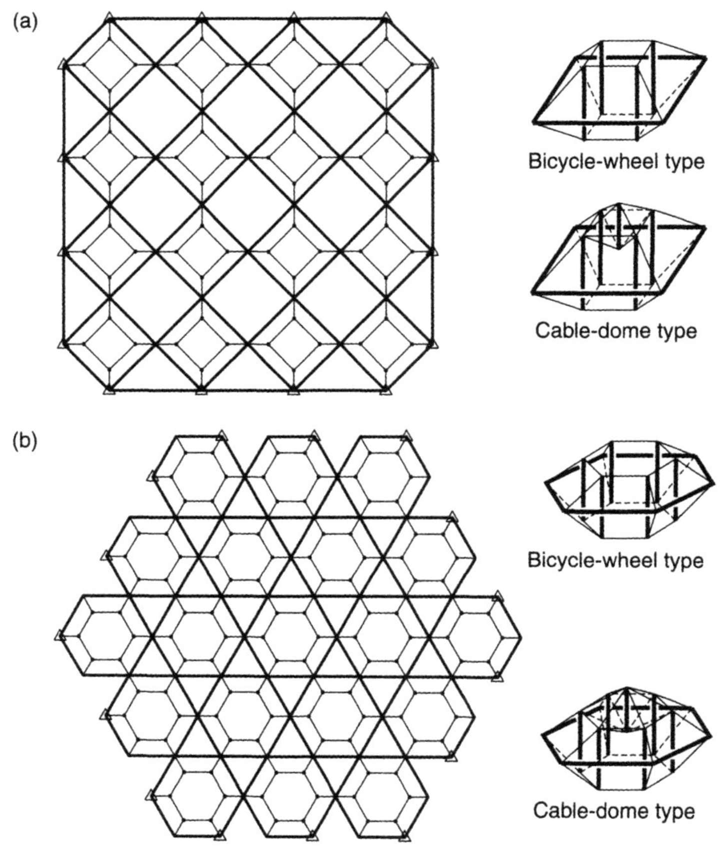

Figure 7.40 'Bicycle-wheel' and 'cable-dome' types: (a) square RP form; (b) hexagonal RP form.

- Struts from the same face are connected to the same inner joint. It forms a 'pyramid' with the face.
- Struts from different faces are not connected to the same inner joint. These inner joints are connected by additional struts.

As the resulting 'pyramidal' bases are isolated, it may be named 'isolated face' method.

The polyhedron in Figure 7.41(a) contains twelve surfaces, among which two triangles and one rectangle are singled out as 'isolated' faces (Figure 7.41(e)). Four struts from the rectangle share a joint inside the volume, three struts from a triangle being joined at another joint. Finally, the resulting three inner joints are linked by a small strut triangle. Furthermore, Figure 7.42(a) shows a truncated tetrahedron in which four triangular faces are 'spaced' by three hexagons. Four inner joints

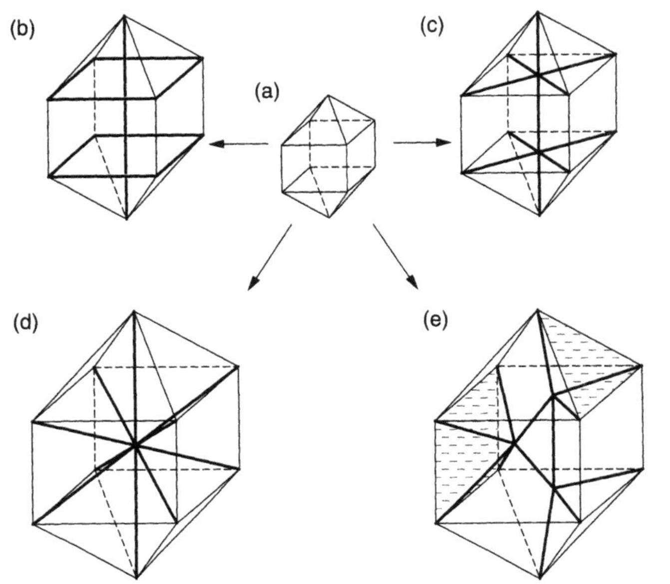

Figure 7.41 RP (DP) derivature and its cable-strut forms; (a) a RP/DP derivature; (b) RP-like type; (c) DP-like type; (d) joined at centre; (e) 'isolated face' method.

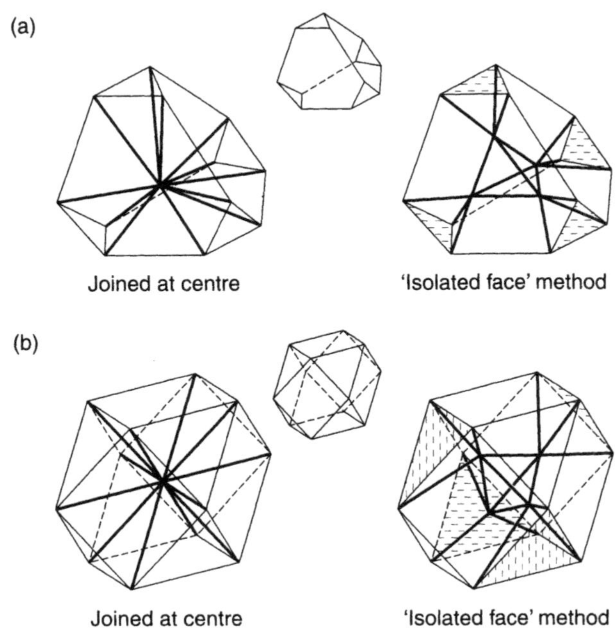

Figure 7.42 Higher cable-strut polyhedra: (a) a truncated tetrahedron; (b) a cub-octahedron.

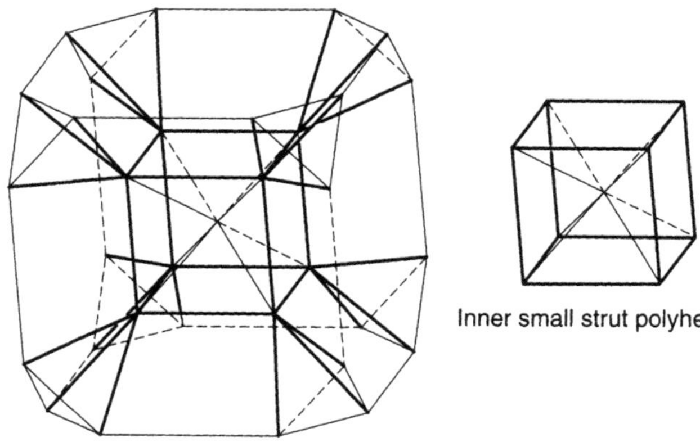

Figure 7.43 Cable-strut truncated cube with small inner strut polyhedra stabilized by cables.

become the vertices of a small strut tetrahedron. The case is similar for the cub-octahedron in Figure 7.42(b).

In higher polyhedra, inner strut polyhedra may not be stable, but it can be stabilized by introducing cables from inside. As an example, the truncated cube is presented in Figure 7.43. The inner cube is stabilized by crossing cables.

7.9 Summary

Cable-strut forms are developed from tensegrity systems and therefore, most morphological studies applicable in tensegrity systems should be suitable for cable-strut forms. In addition, configurations other than 'normal' basic structural forms are proposed (Section 7.3). Special structural and roof design (Section 7.4) and extensively the design with roof sculptures (Section 7.6) are intended to broaden people's imagination. In comparison, ornamental supplements attached to cable-strut simplexes (Section 7.5) are simple. However, many forms can be developed from reshaping basic simplexes (Section 7.7), among which the 'star' group (Section 7.7.1) is especially prominent. Note that irregular simplexes and higher polyhedra (principles presented in Section 7.8) are not focused in the study, but their potential is not negligible.

These are all forms and concepts that the author can propose for the present. *Note that these concepts can also be combined to produce many new forms.* It is of no doubt that new architectural concepts and forms are yet to be developed beyond those which are described in this book when we are constantly seeking inspiration from the nature.

Appendix A
Tensegrity grids in one-way spanning

Generally, tensegrity grids are sensitive to boundary conditions and not advantageous in one-way spanning, as illustrated as follows.

A.1 Non-contiguous tensegrity grids

In one-way spanning cases, parameters are the same set as those in two-way spanning, which is presented in Section 3.3.2. Because in the vertex-to-edge connection (Method Ib) and Layouts B and C in the edge-to-edge connection (Method II), struts are diagonal to the edge in the compressive layer, they are not suitable in one-way spanning in view of concept design.

Layout A, Method II can refer to Figure 3.13, in which the 'square' supports are used for one-way spanning, the results are presented in Table A.1. It shows that deflections are enlarged significantly compared with those in the two-way spanning case (Table 3.1), and self-weight nearly doubles due to high internal forces. It means that non-contiguous strut grids are sensitive

Table A.1 Design results for non-contiguous strut grids of the edge-to-edge connection (Method II) in one-way spanning

Layout	*Prism rotation angle*	*Maximum forces under designed load (10kN)*		*Self-weight (Kg/m^2)*		*Proportion of slack cables*	*Maximum deflection under live load (span)*	*Strut length (m)*
		Compression	*Tension*	*Struts*	*Cables*			
A	45°	77	60	58.0	19.2	924/2880	1/28	4.646
	67.5°	82	56	52.2	17.4	751/2880	1/41	4.529
A+	45°	75	59	52.8	18.7	928/2880	1/28	4.646
Optimal	67.5°	80	54	48.2	16.8	753/2880	1/41	4.529
tube	67.5° +p							
sizes	10[a]	82	56	51.5	18.2	644/2880	1/46	4.529
	20[a]	86	57	54.9	20.2	573/2880	1/53	4.529

Source: Adapted from Wang and Li (2001).

Note

a Prestress value for all struts (unit: 10kN).

to boundary condition. But the proportion of slack cables and the influence of prestress are quite similar to those in two-way spanning.

As the internal forces are high, self-weight optimization can be feasible by introducing tubes of larger outer diameter and smaller wall thickness. These cross-sections are D140t4.5, D159t5, D168t6, D180t6, D194t6, D219t6 and D219t7. The weight reduction is apparent, as shown in Table A.1, and the effect actually increases with span.

A.2 Contiguous strut tensegrity grids

In one-way spanning, parameters are the same as those in two-way spanning (ref. Section 3.3.3). Method II (vertex-and-edge connection) is not optimal for one-way spanning in view of structural design as struts are diagonal to the edge. The grid of Method I (vertex-to-vertex connection), Layout A (67.5°) is presented in Figure 3.14 with 'square' supports for one-way spanning.

The results of Method I, Layout A are presented in Table A.2. Results show that the properties of the 67.5° case are obviously better than the 45° case, especially in stiffness. But it seems unusual that the proportion of cables in slackness in the 45° case is smaller.

Compared with two-way spanning cases (Table 3.2), maximum internal forces and deflections are much increased. But the increment in self-weight is less significant as in non-contiguous strut grids. It shows that contiguous strut grids are less sensitive to boundary conditions.

A.3 Comparison with space truss

Two tensegrity grids are also compared with the SOS grid, in which case the grid depth is prescribed to be 2.5m. The results are presented in Table A.3 for reference. From the comparison, the low structural efficiency of tensegrity grids, especially the non-contiguous strut type, becomes conspicuous.

Table A.2 Design results for contiguous strut grids (Method I, Layout A) in one-way spanning

Prism rotation angle	*Maximum forces under designed load (10kN)*		*Self-weight (Kg/m²)*		*Proportion of slack cables*	*Maximum deflection under live load (span)*	*Strut length (m)*
	Compression	*Tension*	*Struts*	*Cables*			
45°	60	46	26.6	8.1	440/1719	1/52	3.905
67.5°	42	38	21.4	6.1	560/1719	1/100	3.905

Source: Adapted from Wang and Li (2001).

Table A.3 Comparison of tensegrity grids and space truss in one-way spanning

Grid types	*Maximum forces under designed load (10kN)*		*Self-weight (Kg/m^2)*		*Maximum deflection under live load (span)*	*Strut length (m)*	*Tube cross-section (D × t: mm)*	*Cable cross-section (cm^2)*
	Compression	*Tension*	*Bars*	*Cables*				
Non-contiguous strut tensegrity grid	98	70	50.9	21.3	1/39	4.215	140 × 4.5 159 × 5 168 × 6 194 × 6 219 × 6 245 × 7	1, 2, 4, 6, 8, 10, 14
Contiguous strut tensegrity grid	42	38	21.4	6.1	1/100	3.905	89 × 4 114 × 4 140 × 4.5 159 × 5 168 × 6	0.5, 1, 2, 3, 4, 6, 8
Square-on-square offset space grid	18	24	16.3	—	—	3, 3.279	60 × 3.5 76 × 3.75 89 × 4 114 × 4	—

Source: Adapted from Wang and Li (2001).

Appendix B
Design proposal of large span non-contiguous strut tensegrity grids

Non-contiguous strut tensegrity grids, representing cable-strut types in Section 4.4, are subject to structural defects including heavy weight, low stiffness, and large internal forces, etc. However, the form has unique aesthetic value and may meet special architectural requirement. So if the architectural requirement is to realize the grids in large spans, then how to improve their properties? The techniques are discussed incorporating design examples.

B.1 Design examples

The design examples are based on Hanaor's geometrically rigid non-contiguous strut tensegrity grids (Method Ia, triangular simplexes) in 60m span. Sample grid when frequency = 5 is given in Figure B.1(a). The grid is simply-supported and prestress is not introduced into analysis. The following conditions are given for all cases except that modular length (a) is variable with frequency. Note that the tube cross-sections are characteristic of small wall thickness.

Span: 60m
Grid depth (h): 6m
Prism rotation angle in each simplex: 40°
Frequency: 7, 9, 11
Dead load: 30kgf/m^2 (excluding self-weight)
Live load: 70kgf/m^2
Designed strength for tubes: 300MPa
Available tubes for bars: *D*219*t*6, *D*245*t*7, *D*273*t*7, *D*299*t*8, *D*325*t*8, *D*377*t*9, *D*402*t*9, *D*426*t*10, *D*530*t*10
Designed strength for cables: 1,000MPa
Available cross-sections for cables (cm^2): 1, 2, 5, 10, 15, 20, 25, 30

Test studies show that when frequency = 7, very large tensional reactions occur at six corner supports (definition of corner supports refers to Figure B.1(a)), varying between 500kN and 1,200kN. It causes trouble in

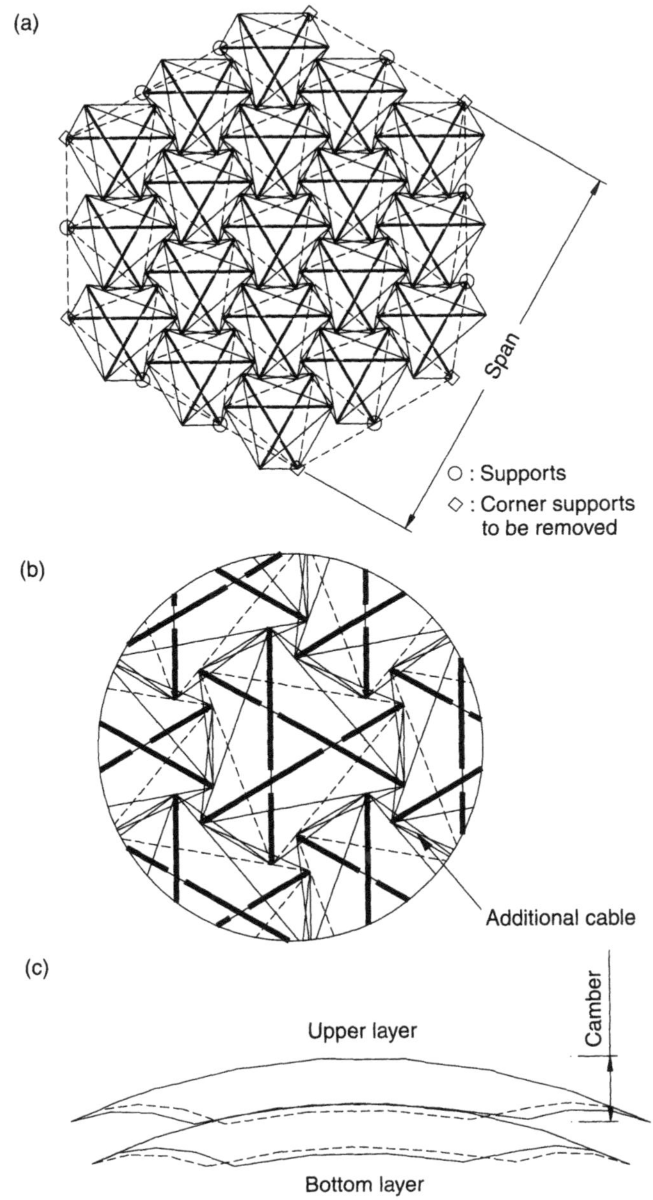

Figure B.1 Non-contiguous strut tensegrity grids made of triangular simplexes: (a) sample planar layout developed by Hanaor (frequency = 5); (b) details of the improved non-contiguous strut configuration; (c) profile of the geometrically rigid configuration with camber.

Table B.1 Results for 60m span non-contiguous strut tensegrity grids

Frequency	*Cases*	*Maximum internal forces (10kN)*		*Self-weight (kg/m²)*		*Maximum deflection under live load (m)*	*Maximum tensional reaction at supports (10kN)*
		Compression	*Tension*	*Struts*	*Cables*		
7	Flat	384	285	38.0	8.3	2.1	11
	Camber	232	225	33.5	6.1	1.4	0
	Improved camber	170	128	31.9	6.1	1.1	0
9	Flat	328	225	42.5	9.9	2.0	60
	Camber	183	165	36.7	7.3	1.4	25
	Improved camber	137	111	34.7	7.3	1.1	22
11	Flat	313	185	48.5	12.1	2.0	90
	Camber	176	139	40.6	9.2	1.4	65
	Improved camber	131	107	39.6	9.2	1.1	48

designing supporting structures. So in the design, six corner supports are released. For grids with higher frequency, more supports tend to be subjected to tensional reactions. But for the convenience of narration, the condition is set the same: only six corner supports are released. Design results are presented in Table B.1. Various factors affecting the design are summarized as follows.

Camber

In all three cases of flat grids (frequency = 7, 9, 11, respectively), internal forces and central deflections are large and support reactions are distributed unevenly due to the mechanism geometry as analysed in Section 3.4.2. When grids with the camber of 10% span are introduced, distributions of support reactions as well as internal forces are much improved, and central deflections are reduced significantly because it is improved to near-mechanism geometry. The self-weight is also reduced but not significant.

Reinforcement in configuration

Properties of grids with camber can be further reinforced by restoring short diagonal cables to each simplex (Figure B.1(b)). Therefore, internal forces are distributed more evenly so that strut weight and deflection can be further reduced (see 'Improved camber' case in Table B.1).

Modular length

The modular lengths for three frequencies are respectively 10.315m, 8.067m, and 6.624m (by contrast, the normal modular length in a space truss for the span is about 4m). Refer to Table B.1 it is obvious that owing to the increase of modular length, the self-weight (excluding joint weight, but it is not much for the large modules) is reduced from 48.8kg/m^2 (frequency = 11), to 42kg/m^2 (frequency = 9), to 38kg/m^2 (frequency = 7). The lightest one is about 50% heavier than space trusses (SOS grid). In addition, the reduction of frequency reduces significantly supports that have tensional reactions (when frequency = 7, no tensional supports remain at the camber cases).

B.2 General principles

General principles of applying non-contiguous strut tensegrity grids in large spans can be summarized as follows.

- Design struts with tubes of larger outer diameter and smaller wall thickness. In non-contiguous strut grids, struts are long, which requires large outer diameter to resist bar buckling, and are subjected to large forces, which requires large cross-section. Designing large tube cross-sections lets cables to look thinner thus the eyesight effect of strut isolation becomes more conspicuous. Note that the feasibility of designing tubes of larger diameter in tensegrity grids is because struts are isolated and joint design can be simple (in conventional bar system, such as space trusses, the introduction of such tubes would increase joint diameters and thus joint weight significantly, consequently, the total weight would increase on the contrary).
- Camber design in place of flat plan. Non-contiguous strut grids are subjected to large deflection. In order to avoid water drainage problem, camber design is necessary. The rise is recommended 10% of the span. Due to the improvement from mechanism geometry to near-mechanism geometry, camber design (of domical shape) has additional advantage in reducing deflections, making even the distribution of internal forces and reducing large tensional reactions occur at supports in flat grids.
- Larger modular length, that is, designing lesser modules, is preferred. When cambered forms have to be employed, the geometry of non-contiguous strut tensegrity grids becomes complex. Benefiting from reduced modules, simplex connection and construction become easier. Structural weight is thus reduced with the additional benefit from the introduction of tubes of larger diameter.
- Reinforcement in configuration. When possible, introducing reinforcing cables (sometimes even bars) can make some further improvement in properties, as the case in Section B.1.

When the modular length or span is further increased, another option is to introduce super strut concept so as to sustain larger forces and to reduce further strut weight, as illustrated in the following section.

B.3 Super strut application

Super strut concept

Under the super bar concept, each long bar is replaced by a reciprocal prism or generally a derivative of reciprocal prism (DRPs) composed of much shorter bars with reduced diameter and wall thickness (Figure B.2). Each segment in the derived part of a DRP is a regular anti-prism, typically a triangular one, so that the axial force can be shared by six inclined bars of equal length. Two super bar ends and all base centres of the derived segments trace a straight line, following the direction of the external axial force. Alternatively, when small cross-section is designed, each segment of

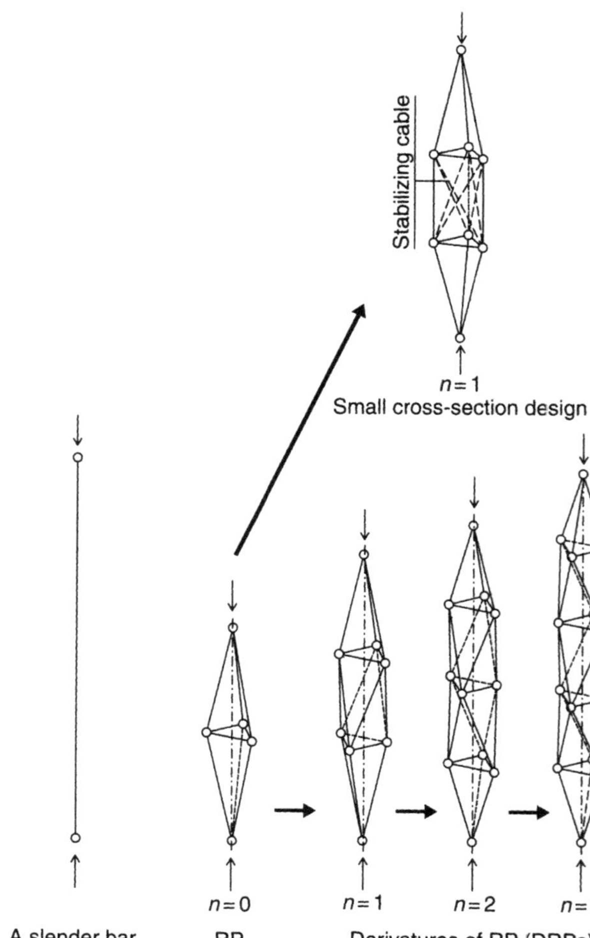

Figure B.2 Concept of super bars.

the derived part is a triangular prism stabilized by crossing cables in lateral faces.

The reciprocal bar prism and designed DRPs are called 'super bars'. The frequency of derivative (n) of a DRP is determined by super bar length, super bar force and dimension optimization of short bars. Super bar system can cope with the stability problem of long struts with reduced weight and large internal forces to joints. It is applied into 200m span space frames (Wang 2000) and 500 m span latticed shells (Wang and Li 1999). Details of super bar design refer to Wang (2000). Super bars can be either tensional or compressive, but are subjected only to compressive forces when introduced into tensegrity grids, thus they are super 'struts'.

Design results of applying super struts

Super struts are introduced into non-contiguous strut tensegrity grids. The parameters and conditions different from those in Section B.1 are given as follows:

Camber: 10%
Frequency: 5
Equivalent super strut cross-sections (cm^2): 60, 80, 100, 120, 140
Cable cross-sections (cm^2): 1, 2, 5, 10, 15, 20.

Owing to the further reduction of frequency to 5, the length of each strut is increased to 17.468m. Note that in this case, reactions in six corner supports are in compression thus these supports are kept. The resulting strut weight is 23kg/m^2, and the cable weight is 4kg/m^2. The total weight including joints is equivalent to that of space trusses (about 30kg/m^2).

Appendix C

Cable-strut grids made of triangular simplexes

Cable-strut grids made of triangular simplexes shall present different appearance from those made of square ones. Conceptually, the ATP grid made of the triangular simplex is not suitable due to excessively high bar density that leads to tensional bars. The P (P2) grid, AP grid, CP grid, RP grid and DP grid are presented in Figures C.1–C.5, respectively. The spans of these grids follow 30m. The depth can refer to the design in the square plan.

In Figure C.1, it is shown a hexagonal plan made of the triangular P (P2), some struts are required at the edge in the compressive layer for stability, and some cables in the tensional layer. The layouts can be easily changed into circular or elliptical ones.

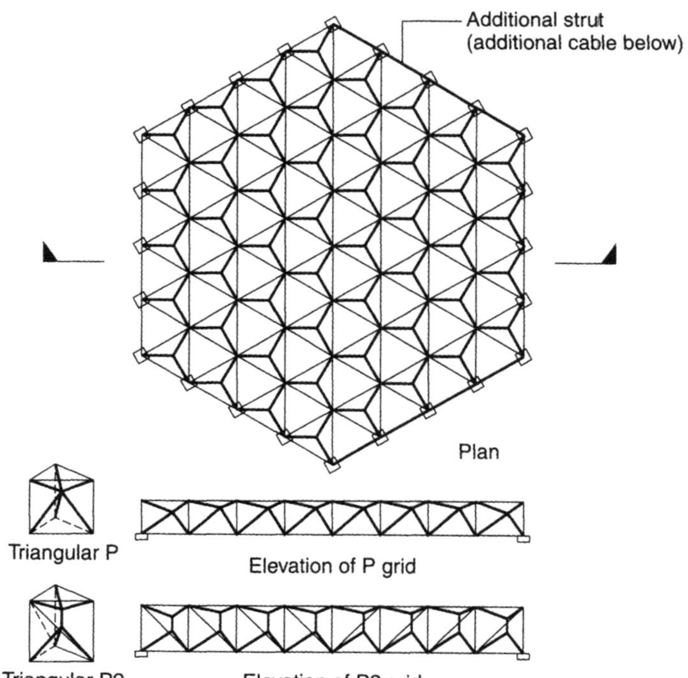

Figure C.1 Sample P (P2) grids made of the triangular simplex.

In Figure C.2, the AP grid, containing seven modules along span, is tailored to suit a circular plan. It presents different appearance in plan from other grids.

In Figure C.3, the CP grid contains eight modules along span, but the modules along circular direction reduce from the periphery, 21, to the

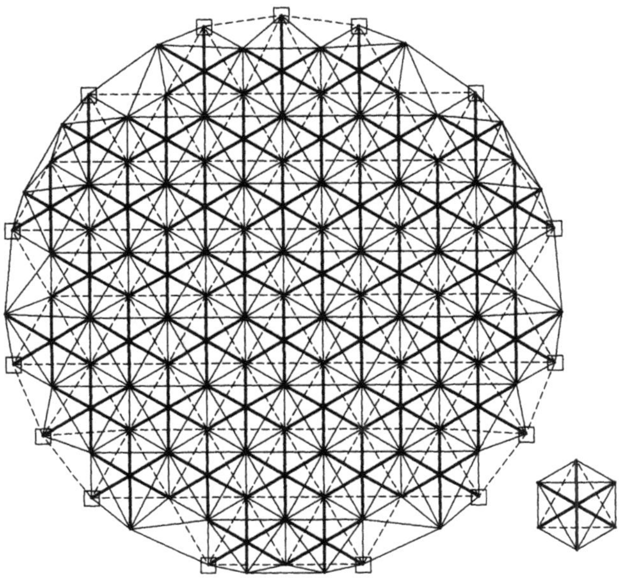

Figure C.2 Sample AP grid made of the triangular simplex.

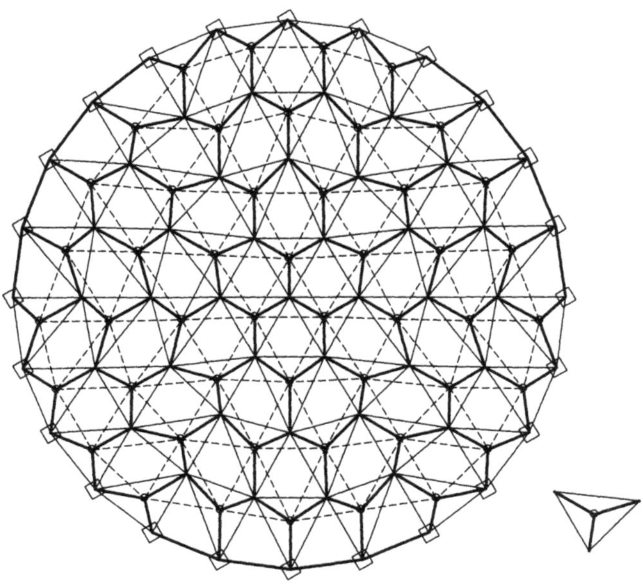

Figure C.3 Sample CP grid made of the triangular simplex.

Figure C.4 Sample RP grid composed of the triangular simplex.

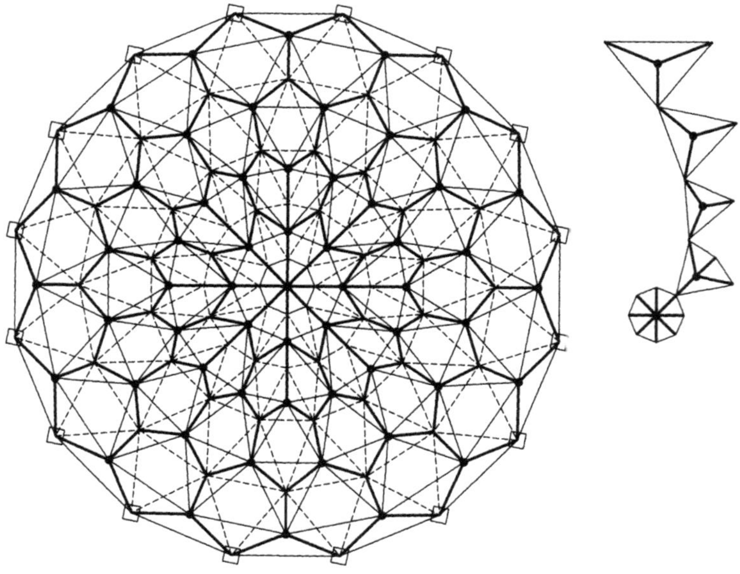

Figure C.5 Sample DP grid composed of the triangular simplex.

centre, three. In Figure C.4, the RP grid is similar to the CP grid but is formed by repeating one-sixth of the grid.

In Figure C.5, the DP grid presents different design. The modules along the circular direction are the same for the outer three rings, followed by a ring of eight quadrilateral DPs and an octagon-based DP core to avoid bar congestion.

In general, Figures C.2–C.5 show different triangulated ways of forming circular grids among a wide variety of choices. Which one is more suitable is related to the specific grid type, and shall be determined by structural analysis.

Appendix D

Cable-strut grids in one-way spanning

Various cable-strut grids are studied in one-way spanning (supports are introduced in two opposite edges only) so as to include general supporting conditions for extensive understanding. No prestress is introduced and conditions and layouts are given the same as in Chapter 5, despite that large modules is preferred along span (e.g. 3.75m × 5m in the DP-b grid and P-b grid). Among cable-strut grids, the RP-b grid, the CP-a grid, the DP-b grid, the ATP grid and the AP grid contain strut diagonal to the edge in the compressive layer, thus they are not preferred in one-way spanning. The left grids are studied and compared with the SOS grid (by the way, the CP truss in Figure 7.15 is also a good choice in one-way spanning but is not studied here). The chief design results are given in Table D.1 and the distribution of internal forces in the compressive and tensional layers of various grids is illustrated in Figures D.1–D.5.

D.1 RP-a grid

In the compressive layer of the RP-a grid, compressions along the span are distributed evenly in high values due to the forces transferred by upper inclined cables (Figure D.1), following the same principle as in the planar

Table D.1 Design results for various cable-strut and space grids in one-way spanning

Grid types	*Parameters in design (m)*			*Max. internal forces (10kN)*		*Weight (kg/m²)*		*Deflection under live load (span)*	*Number of joints/bars/cables*
	a	*h*	h_1 *or* h_u	*Compression*	*Tension*	*Strut*	*Cable*		
RP-a	3	2.5	0.75	23	20	9.3	2.7	1/223	321/320/1160
CP-b	3.75	3	0.5, 1	27	20	5.3	1.8	1/254	272/320/772
DP-b	3.75	2.5	1.5	25	24	6.2	2.0	1/282	336/412/992
P-b	3.75	2.5	0.5, 1	29	34	7.6	2.0	1/314	352/512/1108
SOS	3	2.5	—	20	23	15.4	—	—	221/800/0

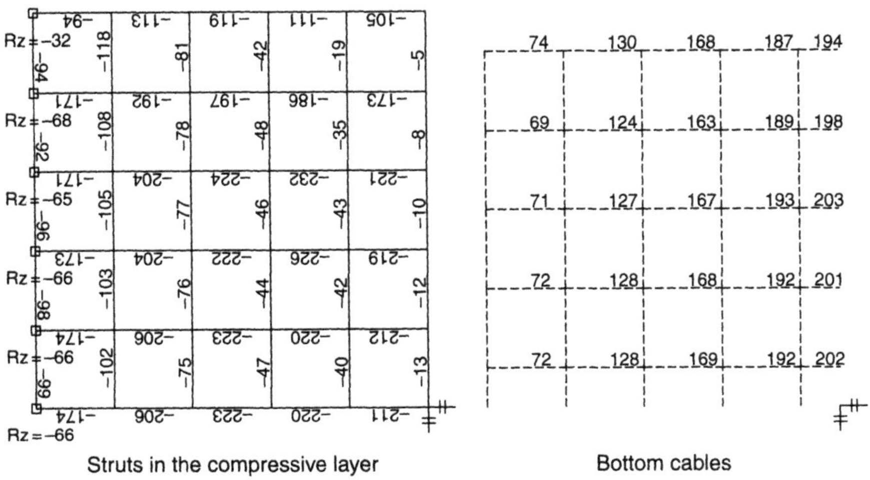

Figure D.1 Internal forces of the RP-a grid in one-way spanning.

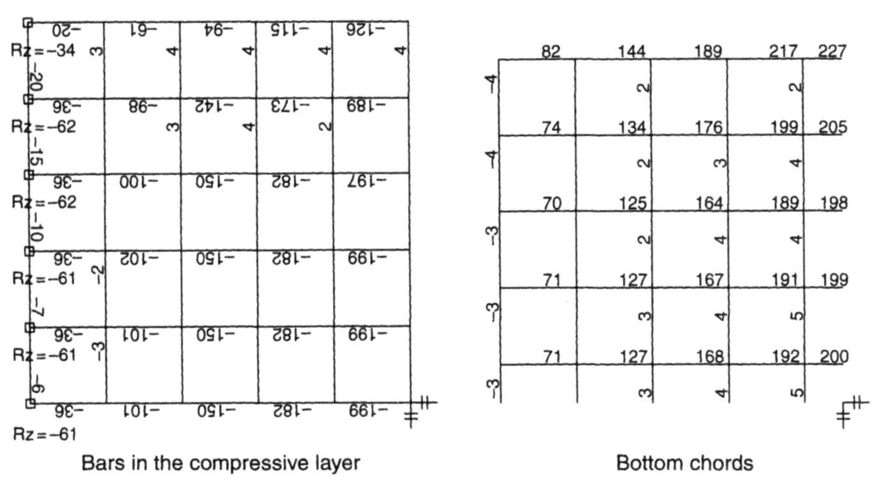

Figure D.2 Internal forces of the SOS grid in one-way spanning.

forms (ref. Section 5.1). Forces in upper inclined cables are also transferred to struts orthogonal to the span. The values are larger near the boundary and smaller near the centre. Tensions in the tensional layer are generally identical to those in the SOS grid (Figure D.2).

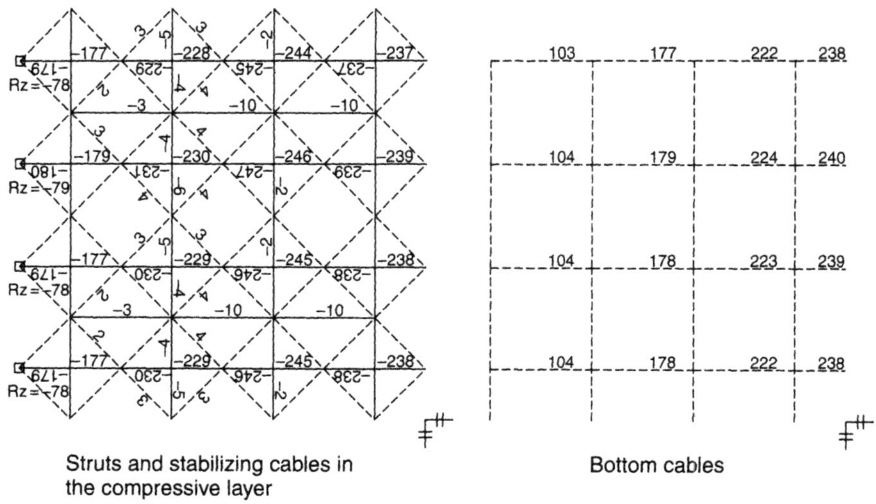

Figure D.3 Internal forces of the DP-b grid in one-way spanning.

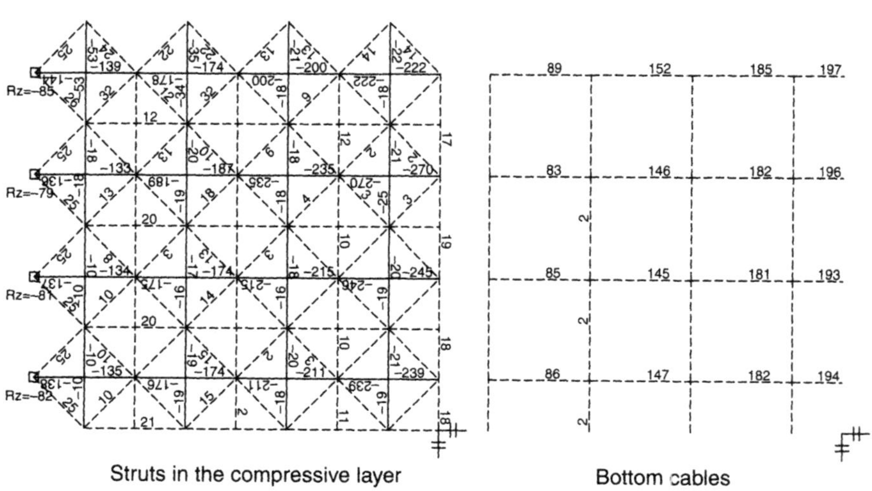

Figure D.4 Internal forces of the CP-b grid in one-way spanning.

D.2 DP-b grid

The distribution of bottom tensional forces and compressions along span in the DP-b grid (Figure D.3) is much similar to that in the RP-a grid and the values are enlarged proportionally with modular length. The difference is that in the DP-b grid, forces in the struts orthogonal to the span are negligible, presenting better load-transfer property and much lighter weight.

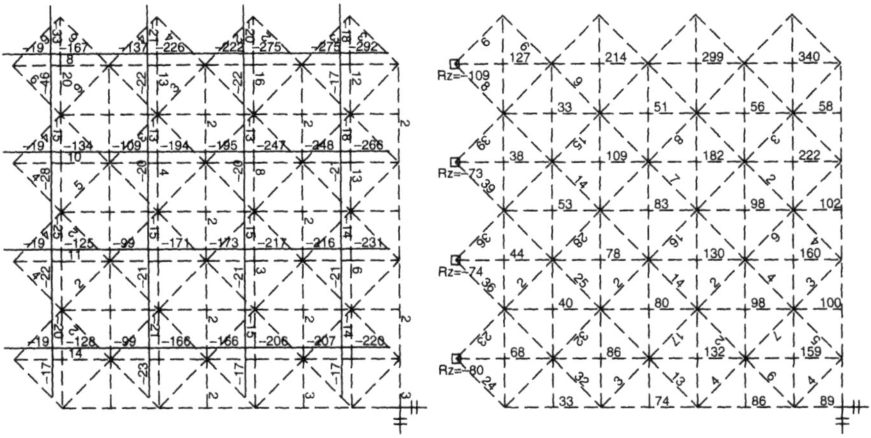

Upper inclined struts and upper cables Bottom cables

Figure D.5 Internal forces of the P-b grid in one-way spanning.

D.3 CP-b grid

In the CP-b grid, its resistant lever arm for bottom cables is 1.25 times that in the SOS grid and the modular length is 1.25 times the latter, so the tensions are approximate to those in the latter (actually smaller due to smaller self-weight load) (Figure D.4). Many upper cables do not slacken but the forces are not large. The resistant lever arm for struts in inner modules, measured approximately by the distance from the bottom joint to the inclined struts in the same module, is 2.416m, 0.8 times that of the bottom cables. Therefore, maximum bar forces shall be around 245kN. The actual distribution of strut compressions is somewhat uneven due to cable slackening but the influence is not much.

D.4 P-b grid

In the P-b grid, the distribution of compressions in upper inclined struts (Figure D.5) is similar as a whole to that in the CP grid. The difference is due to the interaction between the upper and lower inclined struts. In the bottom layer, base cables are subjected to small forces as they are diagonal to the span. Stabilizing cables along span sustain tensions instead. The distribution of internal forces is relatively uneven. However, it is still lighter than the RP-a grid due to the reduction in bar length.

Bibliography

Argyris J.H. and Scharpf D.W. (1972) Large deflection analysis of prestressed networks, *Journal of Structural Division*, ASCE, 98(st3), 633–654.

Berger H. (1997) Shaping lightweight surface structures, *IASS International Symposium'97 on Shell & Spatial Structures*, Singapore, 1, 1–14.

Bouderbala M., Ballouard P., Cevaer F. and Motro R. (1997) Tensegrity systems folding modelization, *IASS International Symposium'97 on Shell & Spatial Structures*, Singapore, 1, 177–185.

Brown L.M. (1998) Case study review of cable suspended roofs, *International Symposium'98 on Lightweight Structures in Architecture, Engineering and Construction*, Sydney, 605–613.

Burt M. (1996) *The Periodical Table of the Polyhedral Universe*, Technion Publication, Israel.

Calladine C.R. (1978) Buckminister Fuller's 'Tensegrity' structures and Clerk Maxwell's rules for the construction of stiff frames, *International Journal of Solids & Structures*, 14, 161–172.

Calladine C.R. and Pellegrino S. (1991) First-order infinitesimal mechanisms, *International Journal of Solids & Structures*, 27, 505–515.

Chassagnoux A., Chomarat S. and Savel J. (1992) A study of morphological characteristics of tensegrity structures, *International Journal of Space Structures*, England, 7(2), 165–172.

Emmerich D.G. (1963) *Construction de Reseaux Autotendants*, French Patent 1,377,290, filed April 10, 1963, granted Sep. 28, 1964.

Emmerich D.G. (1990) Self-tensioning spherical structures: single and double layer spheroids, *International Journal of Space Structures*, England, 5(3 and 4), 351–374.

Emmerich D.G. (1993) Absolute minimal self-tensioning configurations, *Space Structures*, 4, G.A.P. Parke and C.M. Howard, Eds, Thomas Telford, London, 998–1007.

Emmerich D.G. (1996) Emmerich on self-tensioning structures, *International Journal of Space Structures*, England, 11(1 and 2), 29–36.

Escrig F. and Valcarcel J.P. (1993) Geometry of expandable space structures, *International Journal of Space Structures*, England, 8(1 and 2), 71–84.

Escrig F., Valcarcel J.P. and Sanchez J. (1996) Deployable cover as a swimming pool in Seville, *Journal of the International Association for Shell & Spatial Structures*, 37(1), 39–70.

Fanning P. and Hollaway L. (1993) The deployment analysis of a large space antenna, *International Journal of Space Structures*, 8(3), 209–220.

Fuller R.B. (1962) *Tensile Integrity Structures*, US patent 3063521, filed August 31, 1959, granted November 13, 1962.

Fuller R.B. (1975) *Synergetics: Explorations in the Geometry of Thinking*. Collier Macmillan Publishers, London.

Grip R. (1992) The correspondence between convex polyhedra and tensegrity systems: a classification system, *International Journal of Space Structures*, England, 7(2), 115–125.

Hanaor A. (1991a) Double-layer tensegrity grids: static load response, I – analytical study, *Journal of Structural Engineering*, ASCE, 117(6), 1660–1674.

Hanaor A. (1991b) Double-layer tensegrity grids: static load response, II – experimental study, *Journal of Structural Engineering*, ASCE, 117(6), 1675–1684.

Hanaor A. (1992) Aspects of design of double-layer tensegrity domes, *International Journal of Space Structures*, England, 7(2), 101–114.

Hanaor A. (1993) Double-layer tensegrity grids as deployable structures, *International Journal of Space Structures*, England, 8 (1 and 2), 135–145.

Hanaor A. (1994) Geometrically rigid double-layer tensegrity grids, *International Journal of Space Structures*, England, 9(4), 227–238.

Hanaor A. (1997) Tensegrity: theory and application, *Beyond the Cube*, J.F. Gabriel, Ed., John Wiley & Sons, New York, 385–408.

Hanaor A. (2000) Some structural morphological aspects of deployable structures for space enclosures, *IASS Structural Morphology Colloquium 2000 – Bridge Between Civil Engineering and Architecture*, 196–205.

Hanaor A. (2002) The concept of structural depth as applied to certain bar–tendon assemblies, *Proceedings of the Fifth International Conference on Space Structures*, Surrey, 11–20.

Hangai Y. and Lin X.G. (1989) Geometrically non-linear analysis in the vicinity of critical points by the generalized inverse, *International Journal of Space Structures*, England, 4(4), 181–191.

Hangai Y., Kawaguchi K. and Oda K. (1992) Self-equilibrated stress system and structural behaviours of truss structures stabilised by cable tension, *International Journal of Space Structures*, 7(2), 91–99.

Hilyard M.J. and Lalvani H. (2000) Emmerich-type structures constructed from tensegrity modules, *IASS Structural Morphology Colloquium 2000, Bridge Between Civil Engineering & Architecture*, Delft, 196–205.

Hollaway L. and York D. (1995) Numerical analysis of an energy-loaded joint for a deployable satellite structure, *International Journal of Space Structures*, England, 10(1), 47–55.

Hooper P. (1998) Tensile tiles: the cubic modularity of curved surfaces, *Proceedings of LSA98 'Lightweight Structures on Architectural Engineering and Construction'*, Sydney, 2, 1007–1018.

Huybers P. (2002) Frames of nested polyhedra, *Space Structures 5*, G.A.R Parke and P. Disney, Eds, Thomas Telford, Guildford, 449–458.

Imai K., Fujita Y. and Furukawa T. (2002) Development of the KT-wood space truss system with round timber as a new structural material, *Space Structures 5*, G.A.R Parke and P. Disney, Eds, Thomas Telford, Guildford, 155–160.

Ishii K. (1999) *Membrane Designs and Structures in the World*, Shinkenchiku-shu, Tokyo.

Ishii K. (2000) *Structural Design of Retractable Roof Structures*, WIT press, Southampton, UK.

Kawaguchi M., Abe M. and Tatemichi J. (1999) Design, tests and realization of 'suspen-dome' system, *Journal of the International Association for Shells & Spatial Structures*, 40(3), 179–192.

Kono Y., Choong K.K., Shimada T. and Kunieda H. (1999) An experimental investigation of a type of double-layer tensegrity grids, *Journal of the International Association for Shell & Spatial Structures*, 40(2), 103–111.

Lee B.H. (2001) *Analysis, Design and Implementation of Cable-strut Structures*, Master's Thesis, National University of Singapore.

Levy M.P. (1994) The Georgia dome and beyond achieving lightweight – long span structures, *Spatial, Lattice and Tension Structures. Proceedings of the IASS-ASCE International Symposium*, 560–562.

Liapi K. (2000) Tensegrity systems and structural and architectural conceptioning, *IASS Structural Morphology Colloquium 2000 – Bridge Between Civil Engineering and Architecture*, 235–240.

Liddell I. (1998) Theory and practice in the design of lightweight structures, the design of the Millennium Dome, *International Symposium'98 on Lightweight Structures in Architecture, Engineering and Construction*, Sydney, 672–683.

Motro R. (1990) Tensegrity systems and geodesic domes, *International Journal of Space Structures*, England, 5(3 and 4), 341–351.

Motro R. (1996) Structural morphology of tensegrity systems, *International Journal of Space Structures*, England, 11(1 and 2), 233–240.

Motro R. (1999) Special issue on form finding of tension structures, *International Journal of Space Structures*, England, 14(2), 73–154.

Motro R. (2002) Tensarch: a tensegrity double-layer grid prototype, *Space Structures 5*, G.A.R Parke and P. Disney, Eds, Thomas Telford, Guildford, 57–66.

Pavlov G.N. (2002) Methods of virtual architecture used for the design of geodesic domes and multi-petal shells, *Space Structures 5*, G.A.R Parke and P. Disney, Eds, Thomas Telford, Guildford, 673–681.

Pedretti M. (1998) Smart tensegrity structures for the Swiss Expo 2001, *IASS'98 Symposium*, Sydney, 2, 684–691.

Pelligrino S. (1992) A class of tensegrity dome, *International Journal of Space Structures*, England, 7(2), 127–142.

Pilla D., Berger H., Denardes J., Maloney C. and Rois J. (1998) Hybrid cable-arch systems for long span, lightweight roof structures, *International Symposium'98 on Lightweight Structures in Architecture, Engineering and Construction*, Sydney, 133–140.

Pugh A. (1976) *An Introduction to Tensegrity*, University of California Press, Berkeley.

Rice P. and Dutton H. (1995) *Structural Glass*, E&FN Spon, England.

Sadao S. (1996) Fuller on tensegrity, *International Journal of Space Structures*, England, 11(1and2), 37–42.

Saitoh M. (1998) Aesthetics and technology of the beam string structures, *IASS International Symposium'98 on Lightweight Structures in Architecture, Engineering and Construction*, Sydney, 692–701.

Schlaich J. and Schober H. (1999) Recent glass roofs, *Journal of the International Association for Shells & Spatial Structures*, 40(3), 193–205.

Schlaich J., Bergermann R. and Goppert K. (1997) National sports complex – Kuala Lumpur, Malaysia, lightweight roof structures for the outdoor stadium and the swimming complex, *IASS International Symposium'97 on Shell & Spatial Structures*, Singapore, 735–743.

Snelson K. (1996) Snelson on tensegrity, *International Journal of Space Structures*, England, 11(1and2), 43–48.

Wang B.B. (1995) Feasibility studies of tensegrity, Doctoral thesis, Tianjin University, October, in Chinese.

Wang B.B. (1996a) A new type of self-stressed equilibrium cable-strut systems made of reciprocal prisms, *International Journal of Space Structures*, England, 11(4), 349–355.

Wang B.B (1996b) Tensegrity structures as 'ring beams', *Journal of the International Association for Shell & Spatial Structures*, Spain, 37(1), 31–38.

Wang B.B. (1998a) Cable-strut systems: I – tensegrity, *Journal of Constructional Steel Research*, England, 45(3), 281–289.

Wang B.B. (1998b) Cable-strut systems: II – cable-strut, *Journal of Constructional Steel Research*, England, 45(3), 291–299.

Wang B.B. (1998c) Application of the BP system in space structures, *Journal of the International Association for Shell & Spatial Structures*, Madrid, Spain, 14(3), 169–174.

Wang B.B. (2000) Super bar theory and BP system in super span flat space frames, *Journal of Structural Engineering*, ASCE, 126(11), 1365–1370.

Wang B.B. (2002a) *A Type of Cable Grid Structures*, Chinese Patent No. ZL00112025.5, filed January 14, 2000, granted December 11, 2002.

Wang B.B. (2002b) *Load-bearing Cable Grid*, Chinese Patent No. ZL01237975.1, filed May 24, 2001, granted March 6, 2002.

Wang B.B. and Li Y.Y. (1999) A theoretical study of super span latticed shells, *Journal of Constructional Steel Research*, England, 51(3), 287–296.

Wang B.B. and Li Y.Y. (2001) From tensegrity grids to cable-strut grids, *International Journal of Space Structures*, England, 16(4), 279–314.

Wang B.B. and Li Y.Y. (2003a) Novel cable-strut grids made of prisms: I. Basic theory and design, *Journal of the International Association for Shells & Spatial Structures*, 44(2), 93–108.

Wang B.B. and Li Y.Y. (2003b) Novel cable-strut grids made of prisms: II. Deployable and architectural studies, *Journal of the International Association for Shells & Spatial Structures*, 44(2), 109–125.

Wang B.B. and Liu X.L. (1996) Integral-tension research in double-layer tensegrity grids, *International Journal of Space Structures*, England, 11(4), 357–362.

Wang B.B. and Liu X.L. (1998) General study of tensegrity grids of bar-to-bar connection, *International Journal of Space Structures*, England, 13(1), 29–33.

Wang B.B., Xu S.Z. and Liu X.L. (1998) Linear complementary equation method applied in the load response of cable-strut systems, *International Journal of Space Structures*, England, 13(1), 35–40.

Wester T. (2000) The structural morphology of curved diaphragms – or the structural behaviour of floral polyhedra, *IASS Structural Morphology Colloquium 2000, Bridge Between Civil Engineering & Architecture*, 152–159.

Index